103 Advances in Polymer Science

Free Radical Copolimerization Dispersions Glassy State Relaxation

With contributions by
T. S. Chow, L. B. Kandyrin,
S. Kuchanov, V. N. Kuleznev

With 62 Figures and 28 Tables

Springer-Verlag
Berlin Heidelberg GmbH

ISBN 978-3-662-14973-7   ISBN 978-3-540-46727-4 (eBook)
DOI 10.1007/978-3-540-46727-4

Library of Congress Catalog Card Number 61-642

© Springer-Verlag Berlin Heidelberg 1992
Originally published by Springer-Verlag Berlin Heidelberg New York in 1992
Softcover reprint of the hardcover 1st edition 1992

02/3020-5 4 3 2 1 0 — Printed on acid-free paper

Editors

Table of Contents

Modern Aspects of Quantitative Theory of Free-Radical Copolymerization

Semion I. Kuchanov

Polymer Department, Faculty of Chemistry, Lomonosov Moscow State University, Moscow, V-234, 119899, USSR

This review aims to present as completely as possible the description of the current concepts of quantitative radical copolymerization theory. Its various aspects are discussed on the basis of a critical analysis of numerous original publications dealing with this problem. A number of approaches to the solution of the theoretical problems of practical importance, and their application to the prediction of copolymer properties have been considered. Serious attention is also paid to the discussion of kinetic copolymerization models, their experimental corroboration and discrimination of the above models to the description of concrete systems.

1 Introduction

The copolymerization method provides us with the opportunity of producing materials with required properties. Even if we only take such widely used commercial monomers as styrene, vinyl chloride, vinylidene chloride, acrylonitrile, vinyl acetate, low acrylates and methacrylates, the number of different copolymers, including multicomponent ones, which can be synthesized on their basis, is several hundreds. In addition, naturally, samples of copolymers of different compositions and microstructures with appreciably different properties can be obtained from the same set of monomers by changing the monomer feed compositions and the processing conditions. The determination of their properties dependence on the feed stoichiometry of monomers by directly looking into the possible variations in the whole range of compositions is a very laborious experimental task even for terpolymers. As for copolymers consisting of more components, this task seems, in general, impossible in practice. On the other hand, an increase in the number of such components considerably broadens the possibilities of its fitting several different operating requirements simultaneously. Hence, it is obvious how important it is to develop a theoretical approach for finding quantitative correlations between the conditions of copolymer synthesis and its physicochemical and mechanical parameters.

This problem has two fundamental aspects: chemical and physical. The former involves calculations of the copolymer statistical characteristics, while the latter − determination of their relation to the properties. Since we have not so far any theoretical expressions (acceptable for practical application) which are obtained within the framework of a rigorous physical consideration of multicomponent polymer systems at the molecular level, different semiempirical correlation expressions are usually used and these are based on the treatment of the numerous experimental data [1−3]. After we have obtained such relations, only the chemical aspects of the problem remain, i.e. finding dependencies of the statistical characteristics of the molecular structures of copolymers on the conditions of their synthesis.

The above-mentioned statistical characteristics, in contrast to the molecular-weight ones, are independent on the kinetic parameters of initiation and termination reactions and are determined only by the relative reactivity ratios of the propagation reaction. Under the radical copolymerization an important role is assigned to the diffusion factors which control the kinetics of recombination of the polymer radicals. It creates some substantial difficulties in the correct quantitative description of the chain termination reaction, which is not available yet even for homopolymerization despite some semiempirical approaches having been reported [4, 5]. Hence, so far there is no universal kinetic model of radical copolymerization which allows one to calculate reliably the reaction rates and the molecular-weight distribution of the final products. At the same time similar models were developed for the propagation reactions thus making it possible to calculate theoretically the composition, molecular microstructure and composition distribution of copolymers obtained in homogeneous systems.

Together with the traditional statistical problems of calculating "instantaneous" values of the copolymer molecular structural characteristics at the fixed monomer stoichiometry, the dynamic problems of describing the system evolution with conversion also arise since during the process monomer consumption takes place in the reaction system. The practical importance of the revealing of the copolymerization regularities at high conversions is quite clear since the industrial processes of copolymer production are usually carried out up to the high conversions. The systematic description of such a copolymerization from the viewpoint of the dynamic system theory has not been reported yet in literature, except in Ref. [6].

The investigation of the copolymerization dynamics for multicomponent systems in contrast to binary ones becomes a rather complicated problem since the set of the kinetic equations describing the drift of the monomer feed composition with conversion in the latter case has no analytical solution. As for the numerical solutions in the case of the copolymerization of more than three monomers one can speak only about a few particular results [7, 8] based on the simplified equations. A simple constructive algorithm [9] was proposed based on the methods of the theory of graphs, free of the above mentioned shortcomings.

However, even without resorting to numerical calculations one can analyze the general quantitative peculiarities of the reaction mixture evolution with conversion. The theory of the dynamic systems allows one, knowing the values of reactivity ratios, to predict the dependence of the general character of copolymer composition change with conversion on its initial value and also to predict the number of modes and degree of composition inhomogeneity [6]. The latter characteristic is of the vital importance since in a number of cases the broadening of the composition distribution reaches such a degree when due to the thermodynamic incompatibility of the individual copolymer fractions one can observe a microphase separation resulting in the clarity disappearance and frequently in a considerable deterioration of the operation properties of polymeric materials [10]. The possibilities of the theoretical predictions of these properties are discussed in Sect. 7 for some concrete cases.

Even in the first publications concerning the copolymerization theory [11, 12] their authors noticed a certain similarity between the processes of copolymerization and distillation of binary liquid mixtures since both of them are described by the same Lord Rayleigh's equations. The origin of the term "azeotropic copolymerization" comes just from this similarity, when the copolymer composition coincides with monomer feed composition and does not drift with conversion. Many years later the formal similarity in the mathematical description of copolymerization and distillation processes was used again in [13], the authors of which, for the first time, classified the processes of terpolymerization from the viewpoint of their dynamics. The principles on which such a classification for any monomer number m is based are presented in Sect. 5, where there is also demonstrated how these principles can be used for the copolymerization when m = 3 and m = 4.

It is worth mentioning that a set of the types of the dynamic behavior of the system in the case of copolymerization of m monomers is principally wider in comparison with the distillation process of an m-component liquid mixture as it has already been remarked [13]. The reason for this lies in the fact that copolymerization is a non-equilibrium process in contrast to distillation. In a particular case of three-component copolymerization such a possibility is shown

[14, 15] when the monomer feed composition does not approach a certain value but has an oscillating character. Such dynamic behavior which is called "auto-oscillations" is principally not possible in distillation processes. Similar regimes have been already carefully studied both theoretically and experimentally for catalytic and biochemical reactions [16]. The famous reaction of Belousov-Zhabotinski [16] is an example of such oscillations. However the kinetic models resulting in chemical oscillations are usually regarded as being more complicated and less reliable in comparison with the traditional model of radical polymerization [14, 15].

The production of copolymers with homogeneous composition from the monomers of sufficiently different reactivity ratios is considered to be an important task of polymerization. It should be noted that the copolymers produced under the traditional batch process are characterized by such a broad composition distribution that due to the limited component compatibility one can observe a formation of the products with microheterogeneous structure. One of the ways of effectively solving this problem is that copolymerization should be carried out in a continuous stirred ideally-mixed tank reactor where the macromolecules of practically the same composition are produced under a stationary regime. However in addition to such regimes one can also realize auto-oscillating ones [15, 17, 18] under which the composition of copolymer and monomer mixture in the reactor output changes periodically although the monomer feed flows in the reactor input are kept constant, and the process is carried out under isothermal conditions without "gel-effect". Obviously, under such a regime one obtains a copolymer which is characterized by a noticeable composition inhomogeneity. Hence, it is necessary to predict theoretically the conditions under which such kinetic auto-oscillations are realized to avoid them in future practical application. In Sect. 8 different peculiarities of copolymerization in continuous flow systems are discussed as they are of great importance for the development of the technological processes.

In addition to the above mentioned dynamic problems of copolymerization theory this review naturally dwells on more traditional statistical problems of calculation of "instantaneous" composition, parameters of copolymer molecular structure and composition distribution. The manner of presentation of the material based on the formalism of Markov's chains theory allows one to calculate in the uniform way all the above mentioned copolymer characteristics for the different kinetic models by means of elementary arithmetical operations. In Sect. 3 which is devoted to these problems, one can also find a number of original results concerning the statistical description of the copolymers produced through the complex radical mechanism.

The practical value of the quantitative theory of radical copolymerization depends to a great extent on the adequacy of the applied kinetic model to the real systems. Hence, in Sect. 6 we shall discuss the issues of model discrimination and also the problems of reliability and validity of the calculations of the model parameters with an account of the potentialities of the modern experimental techniques.

2 Kinetic Models

2.1 Terminal and Penultimate Models

The most widely used model is based on the classic scheme suggested by Mayo and Lewis [19]:

$$\begin{array}{ll} \sim\bar{M}_1 + M_1 \xrightarrow{k_{11}} \sim\bar{M}_1\bar{M}_1\,, & \sim\bar{M}_1 + M_2 \xrightarrow{k_{12}} \sim\bar{M}_1\bar{M}_2 \\ \sim\bar{M}_2 + M_1 \xrightarrow{k_{21}} \sim\bar{M}_2\bar{M}_1\,, & \sim\bar{M}_2 + M_2 \xrightarrow{k_{22}} \sim\bar{M}_2\bar{M}_2 \end{array} \tag{2.1}$$

according to which the activity of the polymer radical is determined only by the type of its terminal unit. The applicability of this kinetic model at low conversions has been already verified for many hundreds of monomer pairs for which the values of following parameters:

$$r_1 \equiv r_{12} = k_{11}/k_{12}\,, \qquad r_2 \equiv r_{21} = k_{22}/k_{21} \tag{2.2}$$

are presented in Tables [20–24].

Moreover, a whole set of monomers with bulky and polar substitutors is known, the copolymerization of which cannot, be described by the classic scheme (2.1). In this case, in order to calculate the copolymer composition, molecular structure and composition distribution, one should use a penultimate model or the model of complex formation.

The former theory suggests that the reactivity of the polymeric radical is determined by the type of both ultimate and penultimate units. In this case the kinetic scheme of propagation reaction can be presented as follows [25]:

$$\begin{array}{ll} \sim\bar{M}_1\bar{M}_1 + M_1 \xrightarrow{k_{111}} \sim\bar{M}_1\bar{M}_1\bar{M}_1\,, & \sim\bar{M}_1\bar{M}_1 + M_2 \xrightarrow{k_{112}} \sim\bar{M}_1\bar{M}_1\bar{M}_2 \\ \sim\bar{M}_1\bar{M}_2 + M_1 \xrightarrow{k_{121}} \sim\bar{M}_1\bar{M}_2\bar{M}_1\,, & \sim\bar{M}_1\bar{M}_2 + M_2 \xrightarrow{k_{122}} \sim\bar{M}_1\bar{M}_2\bar{M}_2 \\ \sim\bar{M}_2\bar{M}_1 + M_1 \xrightarrow{k_{211}} \sim\bar{M}_2\bar{M}_1\bar{M}_1\,, & \sim\bar{M}_2\bar{M}_1 + M_2 \xrightarrow{k_{212}} \sim\bar{M}_2\bar{M}_1\bar{M}_2 \\ \sim\bar{M}_2\bar{M}_2 + M_1 \xrightarrow{k_{221}} \sim\bar{M}_2\bar{M}_2\bar{M}_1\,, & \sim\bar{M}_2\bar{M}_2 + M_2 \xrightarrow{k_{222}} \sim\bar{M}_2\bar{M}_2\bar{M}_2\,. \end{array} \tag{2.3}$$

This scheme is characterized by the four kinetic parameters:

$$r_1 = \frac{k_{111}}{k_{112}}\,, \qquad r_1' = \frac{k_{211}}{k_{212}}\,, \qquad r_2 = \frac{k_{222}}{k_{221}}\,, \qquad r_2' = \frac{k_{122}}{k_{121}} \tag{2.4}$$

according to the number of the types of kinetically different macroradical ends. Each type of the latter is characterized by its own parameter (2.4) which determines the relative reactivity ratios of the given radical under the monomer addition reaction. The penultimate model (2.3) is transformed into the terminal model when $r_1' = r_1$ and $r_2' = r_2$. If only one from these two equalities holds, the system is described by the three-parameter simplified penultimate model. The number of the parameters can be reduced to two if one of the monomers is incapable of homopolymerization. All the above-mentioned versions of the penultimate model are widely used in practice [26–35].

2.2 Consideration of Complexing

The formation of the donor-acceptor complexes $M_1 \ldots M_2$ between the monomers M_1 and M_2 is regarded as being an additional important factor responsible for the deviations of some certain systems from the classic copolymerization model. Also it should be noted that besides the single monomer entrance into the polymer chain a possibility of the monomer addition in pairs as a complex also exists. The corresponding kinetic scheme of the propagation reaction parallel with reactions (2.1) involves four additional ones [36]:

$$
\begin{aligned}
\sim\bar{M}_1 + M_1 \ldots M_2 &\xrightarrow{k_{11}^*} \sim\bar{M}_1\bar{M}_1\bar{M}_2 \, , \\
\sim\bar{M}_1 + M_2 \ldots M_1 &\xrightarrow{k_{12}^*} \sim\bar{M}_1\bar{M}_2\bar{M}_1 \\
\sim\bar{M}_2 + M_1 \ldots M_2 &\xrightarrow{k_{21}^*} \sim\bar{M}_2\bar{M}_1\bar{M}_2 \, , \\
\sim\bar{M}_2 + M_2 \ldots M_1 &\xrightarrow{k_{22}^*} \sim\bar{M}_2\bar{M}_2\bar{M}_1 \, .
\end{aligned}
\tag{2.5}
$$

This scheme is characterized by the following parameters

$$
r_{11}^* = \frac{k_{11}}{k_{11}^*} \, , \qquad r_{12}^* = \frac{k_{11}}{k_{12}^*} \, , \qquad r_{21}^* = \frac{k_{22}}{k_{21}^*} \, , \qquad r_{22}^* = \frac{k_{22}}{k_{22}^*}
\tag{2.6}
$$

which together with equilibrium constant k complex formation, comprise a complete set of parameters of such a kinetic model.

Up-to-now, a whole range of radical copolymerization processes has been studied for which the formation of the complex between the monomers plays a key role [37, 38]. The concentration M_{12} of $M_1 \ldots M_2$ complex is determined by the nature of the monomers. It may be rather high when one of the monomers is regarded as being a strong donor and the another one is a strong acceptor. As a result one can obtain regularly alternating copolymers or the copolymers the microstructure of which is close to them. It has been substantiated that such copolymers can also be produced in those systems in which the differences between the donor – acceptor properties of the monomers are not so well-pronounced. In this case, one more component, for example the Lewis acid, should be added to the system. The latter forms a complex with one of the monomers thus increasing its acceptor capacity and facilitating a formation of a triple complex including a pair of different monomers and the Lewis acid molecule [39]. The propagation reactions of such a polymerization are described by the kinetic scheme (2.5) which involves a triple complex instead of the binary one. Its concentration essentially increases on addition of the complexing agent to the system. Moreover, there is a certain growth of the tendency of monomer units in macromolecules to alternate, so by increasing the concentration of the complexing agent, one can synthesize the

copolymers close to regularly alternating ones. It allows one to produce copolymers
with the same composition but with a different unit distribution by just varying
the concentration of the complexing agent. Hence it provides a possibility of
tailoring the microstructure of the traditional copolymers and their properties in
the course of synthesis.

In order to describe the propagation reactions under radical copolymerization
in the presence of the complexing agent the kinetic scheme should involve in
addition to reactions (2.5) 16 elementary ones of addition of monomers M_1 and
M_2 to the radicals R_1 and R_2. Moreover, each of the components may exist either
in the free form or binded with a complexing agent. Thus instead of any one
reaction of the kinetic scheme (2.1) there will be four reactions corresponding to
the number of the different pairs of the reaction forms. Consequently, the number
of the kinetic constants also increases 4-fold. For instance, the role of k_{11} in the
kinetic scheme (2.1) in the model (2.5) will play k_{11}^{ff}, k_{11}^{fb}, k_{11}^{bf} and k_{11}^{bb}, where the
first and the second letters in the superscript denote correspondingly the forms
(free or binded) of radical and monomer at the moment of the elementary act.
The complete kinetic scheme of the propagation reaction under complex-radical
copolymerization involves, parallel with the above mentioned 16 kinetic constants,
8 additional rate constants corresponding to the addition reactions of the triple
complex to various radical forms. Parallel with such a set consisting of 24 kinetic
parameters

$$k_{ij}^{ff}, k_{ij}^{fk}, k_{ij}^{kf}, k_{ij}^{kk}, k_{ij}^{*f}, k_{ij}^{*k} \qquad (i, j = 1, 2) \qquad (2.7)$$

the system within the framework of the above model is also characterized by the
four equilibrium constants k_i^M, k_i^R ($i = 1, 2$) of the formation of the binary monomer
or radical complexes and by a similar constant k_{12} of the formation of the triple
complex.

2.3 Other Models

Along with the above mentioned copolymerization models, some other models
were reported which had more limited applicability for the description of the real
processes. For instance, one of the modifications of the terminal model, put forward
by Laowry [39], suggests that the back reactions of monomer elimination from the
growing radical should be involved in the kinetic scheme. These reactions can be
ignored for the vast majority of the monomers usually applied in practice since
under the normal conditions the propagation-depolymerization equilibrium is
shifted noticeably to the polymer formation. Just a few monomers, such as
α-methylstyrene, which are characterized by low values of the ceiling temperature
are regarded as exceptions. The further modification of the penultimate model
(2.3) according to Ham [40] can be carried out when one takes into account the
influence of the more remote units in the macromolecule being compared with
penultimate ones. The authors [41] advanced a more complicated model within
the framework of which one should also account for depolymerization reactions.

Penultimate and similar kinetic models are used nowadays principally for the treatment of the experimental data, obtained from the copolymerization of certain monomers like fumaronitrile or maleic anhydride which are characterized by rather strong steric and polar effects.

An explanation of an anomalous behavior of such systems has also been performed within the framework of the dissociative complex participation of the model [42] based on the assumption that once a complex $M_1 \ldots M_2$ attaches to a radical, it disrupts with the release of a monomer molecule. That leads to a corresponding modification of the reaction scheme (2.5) [43].

A quite different explanation of the deviations from traditional behavior of radical copolymerization has been suggested by Tüdos [44], who has advanced a "hot-radical" theory. In the kinetic scheme corresponding to this theory one should account for the reactions of the monomers with the intermediate ("hot") radicals which have not yet lost the reaction heat stored in the course of the previous reaction step.

2.4 Multicomponent Copolymerization

The majority of the above mentioned kinetic schemes were used for the description of the multicomponent copolymerization of three or more types m of monomers. The generalization of the scheme (2.1) for m = 3 was carried out by Alfred and Goldfinger [45] and for arbitrary m — by Walling and Briggs [46]. In this case the terminal model

$$\sim\bar{M}_i + M_j \xrightarrow{k_{ij}} \sim\bar{M}_i\bar{M}_j, \qquad r_{ij} = k_{ii}/k_{ij} \ (i, j = 1, \ldots, m) \tag{2.8}$$

is completely characterized by a set of $m(m - 1)$ reactivity ratios r_{ij}. The values of these parameters are determined by the copolymerization in the binary systems and consequently one can use the values of r_{ij} presented in Tables [20–24] for the calculations of the multicomponent systems within the framework of the terminal model (2.8).

This important peculiarity, which allows one to determine the kinetic parameters of m-component copolymerization on the basis of the analysis of the experimental data obtained under the copolymerization of $m(m - 1)/2$ monomer pairs vanishes if one uses other kinetic models instead of the terminal one. There are a number of models describing multicomponent systems which account for the influence of the penultimate unit [47], the formation of the binary [48] and triple [49] complexes and also for the depolymerization reactions [50]. However, up to now, all such models have a limited range of application since the current experimental techniques do not allow one to determine correctly a great number of their kinetic parameters.

3 Statistical Problems of Copolymerization Theory

3.1 Copolymer Statistics Within the Framework of Simple Models

With the radical copolymerization during the formation of every individual macromolecule, the monomer concentration in the reaction system is regarded as being practically unchanged. It allows one to calculate the values of the copolymer composition X_1, the probabilities $P(U_k)$ of the different sequences U_k of the monomer units and composition inhomogeneity at a given composition x_1 of the monomer feed mixture, and then to average all these "instantaneous" statistical characteristics taking into account the x_1 change during the process. Such a two-step calculation procedure which solves firstly statistical problems and only then dynamic ones is determined by the very specific features of the radical copolymerization and is independent on the model selection. The latter gives the explicit dependencies of the "instantaneous" statistical characteristics on x_1 and reactivity ratios.

Such dependencies are quite usual for the terminal and penultimate models since in these cases the sequence distributions are described by Markov statistics [51–53, 6]. In the former case this description is carried out by means of the Markov chain, the states S_i of which correspond to the individual monomer units

$$S_1 \sim \bar{M}_1, \qquad S_2 \sim \bar{M}_2 \qquad Q = \begin{pmatrix} v_{11} & v_{12} \\ v_{21} & v_{22} \end{pmatrix} \tag{3.1}$$

The elements v_{ij} of matrix Q of the probabilities for going from state i to state j are expressed through the fraction x_1 of monomer M_1 in the mixture:

$$v_{12} = 1 - v_{11} = \frac{1 - x_1}{1 - x_1 + r_1 x_1}, \qquad v_{21} = 1 - v_{22} = \frac{x_1}{x_1 + r_2(1 - x_1)}. \tag{3.2}$$

In the latter case the states S_i of the Markov chain correspond to the pairs of the monomer units:

$$S_1 \sim \bar{M}_1\bar{M}_1, \qquad S_2 \sim \bar{M}_1\bar{M}_2, \qquad S_3 \sim \bar{M}_2\bar{M}_1, \qquad S_4 \sim \bar{M}_2\bar{M}_2 \tag{3.3}$$

the probabilities of the transition between which are simply expressed by the following matrix:

$$Q = \begin{bmatrix} v_{11} & v_{12} & 0 & 0 \\ 0 & 0 & v_{23} & v_{24} \\ v_{31} & v_{32} & 0 & 0 \\ 0 & 0 & v_{43} & v_{44} \end{bmatrix} \tag{3.4}$$

Non-zero elements of this matrix are equal to:

$$v_{ij} = \frac{r^{(i)}x_1}{1 - x_1 + r^{(i)}x_1} \quad (j = 1, 3), \qquad v_{ij} = \frac{1 - x_1}{1 - x_1 + r^{(i)}x_1} \quad (j = 2, 4). \qquad (3.5)$$

The four parameters $r^{(i)}$ are connected with the reactivity ratios (2.4) through the following relations:

$$r^{(1)} = r_1, \qquad r^{(2)} = 1/r_2', \qquad r^{(3)} = r_1', \qquad r^{(4)} = 1/r_2. \qquad (3.6)$$

When the Markov character of unit sequence distribution in the copolymer is established and the elements of matrix $\mathbf{Q}$ are known, the standard procedure of Markov chain theory allows one to obtain the explicit formulae for all the statistical characteristics of the copolymer fraction obtained at given monomer feed composition x_1 by means of the simple algebraic operations [51–53, 6].

So to estimate the copolymer composition solving the system of the linear algebraic equations:

$$\vec{\pi}\mathbf{Q} = \vec{\pi} \rightarrow \sum_{i=1}^{2} \pi_i v_{ij} = \pi_j, \qquad \sum_{i=1}^{2} \pi_i = 1 \qquad (3.7)$$

one should calculate the components π_1, π_2 of the stationary vector $\vec{\pi}$ of the corresponding Markov chain

$$\pi_i = \frac{\Delta_i}{\Delta} \quad (i = 1, 2), \qquad \Delta = \sum_{i=1}^{2} \Delta_i. \qquad (3.8)$$

By substituting matrix (3.1) into Eq. (3.7) the following formulae can easily be obtained:

$$\Delta_1 = v_{21}, \qquad \Delta_2 = v_{12} \qquad (3.9)$$

which, taking into account formulae (3.2) and (3.8), lead to the well known Mayo-Lewis expressions [19] for the copolymer composition:

$$X_1 \equiv P(\bar{M}_1) = \pi_1, \qquad X_2 \equiv P(\bar{M}_2) = \pi_2 \qquad (3.10)$$

within the framework of the terminal model. In the case of the penultimate model (2.8) the solution of Eq. (3.7) after the substitution of the matrix (3.4) gives the following relations:

$$\Delta_1 = v_{31}v_{43}, \qquad \Delta_2 = \Delta_3 = v_{12}v_{43}, \qquad \Delta_4 = v_{12}v_{24} \qquad (3.11)$$

which according to formulae (3.8) and (3.3) determine the fractions of the different dyads in the copolymer:

$$P(\bar{M}_1\bar{M}_1) = \pi_1, \qquad P(\bar{M}_1\bar{M}_2) = P(\bar{M}_2\bar{M}_1) = \pi_2 = \pi_3, \qquad P(\bar{M}_2\bar{M}_2) = \pi_4.$$
$$(3.12)$$

Then its composition can obviously be estimated by means of the following relations:

$$X_1 \equiv P(\bar{M}_1) = \pi_1 + \pi_2, \qquad X_2 \equiv P(\bar{M}_2) = \pi_3 + \pi_4 \qquad (3.13)$$

which according to expressions (3.5) can easily be transformed into the form reported in [25].

Within the framework of the above models the problem of the calculation of the sequence distribution is solved in a quite simple way [51–53, 6]. In order to find the probability of any sequence $\{U_k\}$ consisting of k units, it should be expressed through the sequence of Markov chain states, the probability of which is calculated usually by means of the routine procedure as a product of the few factors. The first factor π_i corresponds to the initial state S_i, and each of the following factors, v_{ij}, corresponds to the transition from the state S_i to S_j at the conditional movement along the sequence of Markov chain states. For instance, in this manner one can calculate the probability of the sequence $\{U_3\} = \{\bar{M}_1\bar{M}_2\bar{M}_2\}$ in the both cases of terminal model:

$$P\{\bar{M}_1\bar{M}_2\bar{M}_2\} = P\{S_1S_2S_2\} = \pi_1 v_{12} v_{22} \qquad (3.14)$$

and penultimate model:

$$P\{\bar{M}_1\bar{M}_2\bar{M}_2\} = P\{S_2S_4\} = \pi_2 v_{24}. \qquad (3.15)$$

Essentially, the dependencies of π_i and v_{ij} on the monomer feed composition x_1 in the formula (3.14) and (3.15) are quite different since they correspond to the different models.

The composition distribution of the macromolecules produced according to the schemes (2.1) and (2.3) is determined in addition to its mean value X_1 only by the dispersion σ^2 [6]. The latter quantitatively characterizing the width of the composition distribution is reciprocal to the average degree of polymerization of the macromolecules. The proportionality coefficient D called the "index of sequence inhomogeneity" is calculated for Markov copolymers through the routine procedure [6] using the transition probability matrix $\mathbf{Q}$. The above procedure, applied, for example, in the case of the terminal model, yields an expression for $D = v_{12}v_{21}(v_{11} + v_{22})/(v_{12} + v_{21})^3$ which with the account for expressions (3.2) essentially coincides with the expression for D derived through the quite different way [54].

3.2 Copolymer Statistics Within the Framework of the Complex Participation Model

In contrast to the above mentioned models, the similar statistical description of the products of the complex-radical copolymerization occurring through the scheme (2.5) has been carried out quite recently [37, 49, 55–60]. Within the framework of this Seiner-Litt model, both copolymer composition [37, 49, 55–58] and fractions of the different triads and blocks of the monomer units in the macromolecules were calculated [57]. The probability approaches which were applied in these works, are regarded as being of limited applicability in contrast to the general statistical method [49, 59, 60]. By means of the latter method, the sequence distribution and composition inhomogeneity of the copolymer were completely described [49, 60] and also thorough calculations of its microstructure with the account for the tacticity were carried out [59, 60].

The consistent kinetic analysis of the copolymerization with the simultaneous occurrence of the reactions (2.1) and (2.5) leads to the conclusion that the probabilities of the sequences of the monomer units $\bar{M}_1$ and $\bar{M}_2$ in the macromolecules can not be described by a Markov chain of any finite order. Consequently, in this very case we deal with non-Markovian copolymers, the general theory for which is not yet available [6]. However, a comprehensive statistical description of the products of the complex-radical copolymerization within the framework of the Seiner-Litt model via the consideration of the certain auxiliary Markov chain was carried out [49, 59, 60].

This chain is characterized by four states S_1, S_2, S_3, and S_4, for the determination of which we shall distinguish the monomer units by their conditional color: black or white. The unit $\bar{M}_i$ is assumed to be black when the corresponding monomer M_i adds to the radical as a first monomer of the complex. In the other cases when monomer M_i adds individually or as a second monomer of the complex, the monomer unit $\bar{M}_i$ is white. Now, the monomer unit state is characterized by two indications — its type ($i = 1, 2$) and color. For example, we shall speak about the monomer unit being in the state S_1 when this unit is of the first type and white-colored, i.e. $\bar{M}_1^w$. The other states are determined in the similar manner:

$$S_1 \sim \bar{M}_1^w, \qquad S_2 \sim \bar{M}_1^b, \qquad S_3 \sim \bar{M}_2^w, \qquad S_4 \sim \bar{M}_2^b. \qquad (3.16)$$

In the real polymer chains each monomer unit, essentially, does not "remember" the way of its introduction into the macromolecule. It is characterized only by its type and from this viewpoint is regarded as being uncolored. All the experimental characteristics of the copolymer microstructure are described undoubtedly by the sequences of the uncolored units. Hence, it is quite clear that each state of the sequence of the uncolored units $\bar{M}_1$ is the result of the unification of the corresponding pair of the colored units, i.e. $\bar{M}_1 = S_1 + S_2$, $\bar{M}_2 = S_3 + S_4$. The rigorous kinetic consideration within the framework of the scheme (2.1) and (2.5) reveals [49, 60] that the sequences of the conditionally colored units in the macromolecules really form a certain Markov chain. The probabilities of the transition between its states S_i(3.16) are determined by the following matrix:

$$\mathbf{Q} = \begin{bmatrix} v_{11} & v_{12} & v_{13} & v_{14} \\ 0 & 0 & 1 & 0 \\ v_{31} & v_{32} & v_{33} & v_{34} \\ 1 & 0 & 0 & 0 \end{bmatrix} \equiv \begin{pmatrix} \mathbf{Q}_{11} & \mathbf{Q}_{12} \\ \mathbf{Q}_{21} & \mathbf{Q}_{22} \end{pmatrix}. \tag{3.17}$$

The elements of this matrix which differ from zero and unity are presented below:

$$v_{1j} = d_{1j}/d_1, \qquad v_{3j} = d_{3j}/d_3 \quad (j = 1, 2, 3, 4)$$

$$d_{11} = M_1, \qquad d_{12} = a_{11}^* M_{12}, \qquad d_{13} = a_{12} M_2, \qquad d_{14} = a_{12}^* M_{12},$$

$$d_{31} = a_{21} M_1, \qquad d_{32} = a_{21}^* M_{12}, \qquad d_{33} = M_2, \qquad d_{34} = a_{22}^* M_{12}, \tag{3.18}$$

$$d_i = d_{i1} + d_{i2} + d_{i3} + d_{i4} \quad (i = 1, 3); \qquad a_{ij} = 1/r_{ij}, \qquad a_{ij}^* = 1/r_{ij}^*.$$

If we express the complex concentration M_{12} from the equilibrium conditions $M_{12} = kM_1M_2$ through M_1 and M_2 and then substitute it into relations (3.18), we shall obtain the final formulae which allow one to describe completely Markov statistics of the colored units at given monomer concentrations and reactivity ratios (2.2) and (2.6).

For instance, the components $X_i(i = 1, 2)$ of the composition vector $\mathbf{X}$ according to determination (3.16) are connected with components $\pi_i(i = 1, 2, 3, 4)$ of the stationary vector of the auxiliary Markov chain in the same manner (3.13) as in the penultimate model. However, the values of Δ_i in expressions (3.8) are determined, instead of formula (3.11), by the following expressions:

$$\Delta_1 = v_{31} + v_{34}, \qquad \Delta_2 = v_{13}v_{32} + v_{12}(1 - v_{33}),$$

$$\Delta_3 = v_{12} + v_{13}, \qquad \Delta_4 = v_{14}v_{31} + v_{34}(1 - v_{11}). \tag{3.19}$$

The substitution of the transition probabilities (3.18) into these relationships results in a well-known expression derived quite differently by Seiner and Litt [37]. This expression connects the instantaneous copolymer composition with the current concentrations of the monomers and their complex.

Knowing π_i (3.8) and (3.19), we can as an example write down the formulae for the probabilities of the different dyads of the monomer units:

$$P\{\bar{M}_1\bar{M}_1\} = \pi_1(v_{11} + v_{12}), \qquad P\{\bar{M}_1\bar{M}_2\} = \pi_1(v_{13} + v_{14}) + \pi_2,$$

$$P\{\bar{M}_2\bar{M}_1\} = \pi_3(v_{31}v_{32}) + \pi_4, \quad P\{\bar{M}_2\bar{M}_2\} = \pi_3(v_{33} + v_{34}) \tag{3.20}$$

which are derived on the basis of simple statistical considerations. So, the dyad $\{\bar{M}_1\bar{M}_1\}$ is produced when during the conditional movement along the copolymer chain, the first unit is in the state S_1 (the probability of which equals π_1) and at the next step a transition from S_1 to S_1 or S_2 takes place (the respective probabilities are equal to v_{11} and v_{12}). The dyad $\{\bar{M}_1\bar{M}_2\}$ can be formed in three ways: when the first unit is in the state S_1 and then goes to S_3 or S_4; when the first unit is in the state S_2, then inevitably M_2 will be the next unit. The remaining two relationships (3.20) are derived in a similar manner. In Refs. [49, 60] a general algorithm was proposed that allows, without such tedious considerations, to put down at once the probability of any sequence of the uncolored monomer units as it is used to be done for the terminal model. To carry out this procedure one should modify the formulae for the probabilities $P\{U_k\}$ of the terminal model in the following way: (a) to substitute π_1 and π_2 for vectors $\vec{\pi}^{(1)}(\pi_1, \pi_2)$ and $\vec{\pi}^{(2)}(\pi_3, \pi_4)$, respectively; (b) each of the scalar factors v_{ij} is to be substituted for the submatrix Q_{ij} of the matrix of the transition probabilities Q (3.17); (c) instead of the last factor 1, one should introduce the column-vector $\vec{1}^t$ which is the transpose of row-vector $\vec{1}(1, 1)$. For instance, the triad probability which within the Mayo-Lewis model is described by formula (3.14), in the framework of the Siener-Litt model is equal to:

$$P\{\bar{M}_1\bar{M}_2\bar{M}_2\} = \vec{\pi}^{(1)}Q_{12}Q_{22}\vec{1}^t = (\pi_1, \pi_2)\begin{pmatrix} v_{13} & v_{14} \\ 1 & 0 \end{pmatrix}\begin{pmatrix} v_{33} & v_{34} \\ 0 & 0 \end{pmatrix}\begin{pmatrix} 1 \\ 1 \end{pmatrix}$$

$$= (\pi_1 v_{13} + \pi_2)(v_{33} + v_{34}). \tag{3.21}$$

It was shown [49, 60] that the composition distribution within the framework of both models mentioned are described by the same formulae. However, the values of the parameters of this distribution — mean copolymer composition and index of sequence homogeneity D — are estimated for each of the models essentially in different ways. So in the case of the complex-radical mechanism, the first of these parameters is derived from relations (3.8) and (3.19), and the latter one is equal to the sum of indices of sequential homogeneity in the chain of the colored units. These indices for such a Markov chain are calculated in the usual manner [6].

The above theory discussed within the framework of the Seiner-Litt model can also be effectively used for the calculations of the statistical characteristics of the products of the copolymerization which proceeds in the presence of the added complexing agent. Its concentration together with the kinetic parameters (2.7) and the equilibrium constants of the formation of the different complexes, as may be concluded from the analysis of the corresponding kinetic scheme, will determine the effective values of the reactivity ratios r_{ij}^e and r_{ij}^{*e}. One should substitute these values instead of parameters (2.2) and (2.6) into expressions (3.18) for the transition probabilities which will then describe the Markov chain of colored units of the copolymers obtained in the presence of the complexing agent. Hence, taking account of the complexing agent in the above theory means a trivial change of the real copolymerization constants for the effective ones.

3.3 Different Methods for the Statistical Description of Binary Copolymers

The most generally used mode of the quantitative description of the copolymer chain structure consists in the indication of the fractions $P(U_k)$ of the different sequences U_k with a small number of their monomer units k. The chain structure may be characterized also by the distributions $f^{(i)}(n)$ of monomer units $\bar{M}_i$ runs for their length n. The statistical moments of functions $f^{(i)}(n)$ determine the average lengths of these runs. Additionally, for the quantitative description of the binary copolymer macromolecule microstructure some other characteristics were proposed [61–71]; several of them and their numerical values for some concrete systems were presented by Tosi [72]. He and his co-workers proposed [73–77] to characterize the randomness of the sequence distribution in terms of "informational entropy". This parameter in contrast to the other similar ones has a general meaning and is regarded as being a real measure of the randomness of any copolymers.

For practical purposes, it is convenient to use a graphical representation of the data on the composition and copolymer chain structure, and an application of the nomograms which allow one to determine the values of the proper parameter in a quick and simple manner. The advantages of such approaches have been already demonstrated [78–82].

All the statistical characteristics of copolymer chain structure and composition inhomogeneity, (including the ones reported in the above papers) can be easily calculated by means of the Markov chain formalism for any of kinetic models presented in Sect. 2. Then it does not seem advisable for the solution of such problems to apply the Monte-Carlo method with which the simulation of the copolymer chain growth was carried out [83–93].

In reality, such a simulation is not necessary even for the determination of the validity limits of the above statistical approach to the quantitative description of the copolymers which is based on the formalism of the stationary Markov chain theory.

The Markovian character of the sequence distribution statistics in the macromolecules results [6, 94] from assumption about the steady-state of the radical concentrations, which usually holds with a high degree of accuracy in the copolymerization processes [6, 95]. It is worth mentioning that along with such kinetic stationarity one should usually speak about the statistical stationarity. It means that when the number of the units in copolymer molecules exceeds 10–15, their composition practically becomes independent on degree of polymerization and is indistinguishable from the value predicted by the stationary Markov chain theory. This conclusion is supported by the theoretical [96, 97, 6] and experimental [98] evidence.

4 Statistical Description of the Multicomponent Copolymers

4.1 Sequence Distribution in Macromolecules

A generalization of the theory of the binary copolymerization for multicomponent systems in the case of the terminal model (2.8) is not difficult since the copolymer microstructure is still described by the Markov chain with states S_i corresponding to the monomer units $\bar{M}_i$. The number m of their types determines the order of the matrix $\mathbf{Q}$ of the transition probabilities, the elements of which can be presented in a simple way [53, 6]:

$$v_{ij} = a_{ij}x_j / \sum_{k=1}^{m} a_{ik}x_k \quad (i, j = 1, 2, ..., m) \sum_{i=1}^{m} x_i = 1 . \tag{4.1}$$

The parameters $a_{ij} \equiv 1/r_{ij}$, the number of which equals $m(m-1)$, are reciprocal reactivity ratios (2.8) of binary copolymers. Markov chain theory allows one, without any trouble, to calculate at any m, all the necessary statistical characteristics of the copolymers, which are formed at given composition $\vec{x}$ of the monomer feed mixture. For instance, the instantaneous composition of the multicomponent copolymer is still determined by means of formulae (3.7) and (3.8), the sums which now contain m items. In the general case the problems of the calculation of the instantaneous values of sequence distribution and composition distribution of the Markov multicomponent copolymers were also solved [53, 6]. The availability of the simple algebraic expressions puts in question the expediency of the application of the Monte-Carlo method, which was used in the case of terpolymerization [85, 99–103], for the calculations of the above statistical characteristics. Actually, the probability of any sequence $\{\bar{M}_i\bar{M}_j\bar{M}_k \ldots \bar{M}_r\bar{M}_s\}$ of consecutive monomer units, selected randomly from a polymer chain is calculated by means of the elementary formula:

$$P\{\bar{M}_i\bar{M}_j\bar{M}_k \ldots \bar{M}_r\bar{M}_s\} = \pi_i v_{ij} v_{jk} \ldots v_{rs} . \tag{4.2}$$

For the quantitative description of the sequence distribution in the multicomponent copolymers the statistical characteristics similar to the ones applied for the description of the binary copolymerization products were used [45, 104–110]. In their well-known paper [45] Alfrey and Goldfinger stated an exponential character of the run distribution $f^{(i)}(n)$ for length n with copolymerization of any number m of monomer types. Besides this distribution and its statistical moments [104–107] other parameters as "alternation degree" were introduced [108, 107], which equals the overall fraction of all heterotriads $P(\bar{M}_i\bar{M}_j)$ $(i \neq j)$, and also the parameter with the similar meaning called "alternating order" [109]. Tosi [110] suggested to use informational entropy as a quantitative measure of the randomness of the multicomponent copolymers:

$$h = - \sum_{j=1}^{m} \sum_{i=1}^{m} \pi_i v_{ij} \log_m v_{ij} . \tag{4.3}$$

The value of h is equal to unity when the copolymerization product is a completely random equimolar copolymer described by Bernulli statistics $v_{ij} = \pi_j = 1/m$, and is equal to zero for any copolymer with fixed sequence alternation, as, for instance, in biological macromolecules.

4.2 "Instantaneous" Chemical Composition Distribution

The problem of the calculation of the instantaneous composition distribution of Markov copolymers consisting of arbitrary number m of monomer units is solved rather easily [6]. Each macromolecule is characterized by the overall number of its units 1 (size) and composition vector $\vec{\zeta}$, the components $\zeta_1, \zeta_2, ..., \zeta_i, ..., \zeta_m$ of which are equal to the fractions of the monomer units $\bar{M}_1, \bar{M}_2, ..., \bar{M}_i, ..., \bar{M}_m$ in this macromolecule. The joint size and composition distribution $f(l, \vec{\zeta})$ (SCD) can be represented as a the product of two factors

$$f(l, \zeta) = f(l) \, W(l \mid \zeta) \tag{4.4}$$

the former of which is regarded as being a distribution of the chains for their degree of polymerization and the latter one — a conditional composition distribution of the macromolecule fraction with fixed value of l. Hence dependence $f(l)$ is described, as in the case of homopolymerization, by the exponential (Flory) distribution, having the mean number of units $\bar{l}$ in one macromolecule as a parameter. The value of $\bar{l}$ can be either calculated theoretically taking into account the initiation and termination reactions in the kinetic scheme or obtained experimentally. The consideration of the mentioned reactions is not needed at all for the determination of the conditional distribution $W(l \mid \vec{\zeta})$, which within the framework of the terminal model (2.8), depends only upon the values of the reactivity ratios r_{ij} and is described by the Gauss formula:

$$W(l \mid \vec{\zeta}) = \frac{1}{\sqrt{(2\pi)^{m-1} |\lambda|}} \exp\left[-\frac{1}{2} \sum_{i,j=1}^{m-1} \Lambda_{ij}(\zeta_i - \pi_i)(\zeta_j - \pi_j) \right]. \tag{4.5}$$

The vector of the mean composition $\vec{\pi}$ and matrix of the moments λ are the parameters of this normal distribution. After the calculation of its determinant $|\lambda|$ and the elements Λ_{ij} of the matrix $\Lambda = \lambda^{(-1)}$ inversed to λ one can, by integrating the distribution (4.5) over any composition region $\vec{\zeta}$, determine the fraction of the copolymer molecules of such compositions. The matrix elements $\lambda_{ij} = D_{ij}/l$ are reciprocal to 1 and proportionality coefficients D_{ij} by means of the following relations:

$$D_{ij} = \pi_i Z_{ij} + \pi_j Z_{ji} - \pi_i \delta_{ij} - \pi_i \pi_j \qquad (i, j = 1, 2, ..., m-1) \tag{4.6}$$

are expressed in terms of the elements Z_{ij} of the fundamental matrix $\mathbf{Z}$ of the Markov chain corresponding to the given copolymer. Its matrix of transition probabilities $\mathbf{Q}$ with elements v_{ij} (4.1) determines the value of $\mathbf{Z}$ through the following formula:

$$\mathbf{Z} = (\mathbf{E} - \mathbf{Q} + \mathbf{Q}_{lim})^{(-1)}, \qquad \mathbf{Q}_{lim} \equiv \lim_{n \to \infty} \mathbf{Q}^n \tag{4.7}$$

where $\mathbf{Q}_{lim}$ is the limit matrix, each row of which is the stationary vector $\vec{\pi}$, and $\mathbf{E}$ is the unit matrix with elements δ_{ij} which are equal to 1 when $i = j$ and to 0 when $i \neq j$. Hence, by means of the simple algebraic operations (the most difficult of which is the calculation of determinants of the m-order which is needed for the calculation of the inverse matrices) one can easily calculate the parameters of normal distribution (4.5) together with its statistical moments. In particular, the most important of them — the moments of the second order — determine dispersions σ_i^2 and coefficients of correlation ϱ_{ij}:

$$
\begin{aligned}
\sigma_i &\equiv \overline{(\zeta_i - \pi_i)^2} = \lambda_{ii}, \\
\varrho_{ij} &\equiv \overline{(\zeta_i - \pi_i)(\zeta_j - \pi_j)}/\sigma_i\sigma_j = \lambda_{ij}/(\lambda_{ii}\lambda_{jj})^{1/2}.
\end{aligned} \tag{4.8}
$$

It is worth noting that the above-mentioned expressions (4.5–4.7) contain, as particular cases, the results obtained both for binary [54] and ternary [112] copolymerization. However, the general formulae (4.6) and (4.7) for indexes of sequential homogeneity of multicomponent copolymers with any m were not obtained earlier by the author of Refs. [111, 113], who investigated this problem theoretically. The approaches applied in the above papers result in cumbersome formulae and are not needed since Eqs. (4.6) and (4.7) can be immediately obtained [6] from the Markov chain theory.

In order to calculate the weight-composition distribution one should substitute in expression (4.4) f(l) for the weight Flory distribution and then integrate the modified expression over all l employing formula (4.5) for $W(l \mid \vec{\zeta})$. In this case, as demonstrated by the authors of Ref. [54], it is very convenient to use instead of ζ_i the natural variables $z_i \equiv (\zeta_i - \pi_i)(\bar{l}/2 \mid \mathbf{D})^{1/2}$, characterizing the deviation of the individual molecule composition $\vec{\zeta}$ from its overall value $\vec{\pi}$. In terms of such variables, the function of the composition distribution is the following:

$$
\frac{|\mathbf{D}|^{\frac{m-2}{2}} \Gamma\left(\dfrac{m+3}{2}\right)}{\pi^{\frac{m-1}{2}}} \left\{ \frac{\lambda}{[1 + S(\bar{z})]^{\frac{m+3}{2}}} + \frac{(1 - \lambda)(m + 3)}{4[1 + S(\bar{z})]^{\frac{m+5}{2}}} \right\} \tag{4.9}
$$

where

$$
S(\bar{z}) = \sum_{i=1}^{m-1} \sum_{j=1}^{m-1} \mathbf{D}^{ij} z_i z_j
$$

where $|\mathbf{D}|$ — the determinant of the matrix (4.6), and $\mathbf{D}^{ij}$ — its cofactors which are equal within the accuracy of $(-1)^{i+j}$ to the values of determinants of $(m - 2)$ order, which are obtained from $|\mathbf{D}|$ by striking out in it row i and colum j. The parameters λ and $1 - \lambda$ in expression (4.9) are equal to the fractions of radicals terminating respectively by addition and recombination mechanisms, and numerical factor $\Gamma((m + 3)/2)$ is the Euler's gamma-function, which equals $(\pi)^{1/2}(m + 1)!/2^{m+1}(m/2)!$ and $[(m + 1)/2]!$ respectively for odd and even values of m. In the case of terpolymerization the lines on which the distribution (4.9) keeps a constant value, are found to form an ellipsis $S(\bar{z}) = const$ at the plane (z_1, z_2). For the copolymerization of four monomers in the three-dimensional space of variables (z_1, z_2, z_3), the equicompositional surfaces form ellipsoids $S(\bar{z}) = Const$.

4.3 A Simple Procedure Available for the Calculation of the Multicomponent Copolymer Composition

Formulae (4.2), (4.3), and (4.5)–(4.7) involve components $\pi_i = \Delta_i/\Delta$ of vector $\vec{\pi}$ of instantaneous copolymer composition (3.8), for calculation of which Walling and Briggs [46] according to the kinetic scheme (2.8) derived the following expressions:

$$\Delta_i = x_i B_i \sigma_i, \qquad \sigma_i(\vec{x}) = \sum_{j=1}^{m} a_{ij} x_j \qquad (i = 1, 2, \ldots, m) \tag{4.10}$$

where the expressions for $B_i(\vec{x})$ are presented as certain determinants of $(m - 1)$ order. These determinants, however, have a cumbersome unattractive form, from which one can hardly understand the character of the dependence of multicomponent copolymer composition $\vec{\pi}$ on monomer feed composition $\vec{x}$ and reactivity ratios $r_{ij} = 1/a_{ij}$. It concerns to a greater extent to Ref. [114]. The author of Ref. [9] proposed a simple algorithm which allows one to obtain explicit expressions for the components of vector $\vec{\pi}$ just from formulae (3.7).

Let us explain this algorithm considering the case of the terpolymerization, for which the following expressions

$$x_1 B_1 = a_{21} a_{31} x_1^2 + a_{21} a_{32} x_1 x_2 + a_{23} a_{31} x_1 x_3,$$
$$x_2 B_2 = a_{12} a_{31} x_1 x_2 + a_{12} a_{32} x_2^2 + a_{13} a_{32} x_2 x_3, \tag{4.11}$$
$$x_3 B_3 = a_{13} a_{21} x_1 x_3 + a_{12} a_{23} x_2 x_3 + a_{13} a_{23} x_3^2.$$

were obtained [45]. Each of the items in the right hand parts of these formulae may be related unequivocally with the proper graph, the sets of which are shown in Fig. 1 in the same order as the items in expressions (4.11). An arrow (arc) of directed graphs (digraphs) presented in Fig. 1, which connects point i with point j, corresponds to each of the factors $a_{ij} x_j$ of formulae (4.11). Accounting for such a correspondence, vice versa, knowing the sets of graphs, presented in Fig. 1, one can directly write expressions (4.11). The procedure is still the same even when considering the copolymers for $m > 3$. Hence, to obtain the formulae similar to

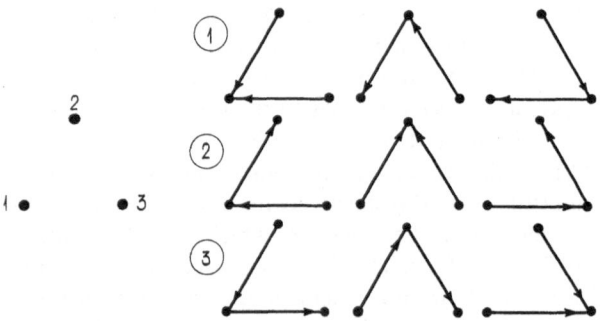

Fig. 1. Three sets of digraphs corresponding to each of expressions (4.11), respectively

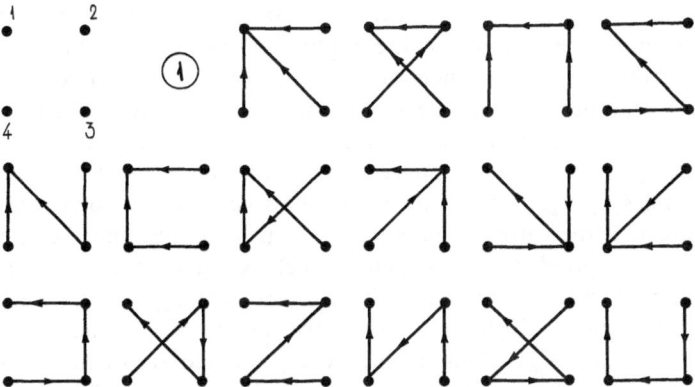

Fig. 2. A complete set of directed trees converging to apex 1

(4.11) ones at any m one should only draw the proper sets of digraphs. It was shown [9] that each i-th of these m sets, corresponding to its own component π_i of copolymer composition vector, involves all converging to point i direct trees containing m points with root in point i. These trees (arborescences) are uncycled graphs, any point of which is connected by a single sequence of arcs (route) similarly directed to the point with number i [115], If all $a_{ij} \neq 0$, according to the well-known theorem of graph theory [115], the number of such directed trees in each of m sets is equal to m^{m-2}.

Any i-th set, if you inverse in it the numbers of points i and j, can be transformed into any other j-th set. This can be easily seen, in particular, in Fig. 1. Hence it is sufficient to draw only one set of direct trees converging, for example, to point i = 1 from which the expression for $x_1 B_1$ can be derived.

Let us consider, for instance, four-component copolymerization. According to the above algorithm each of $4^2 = 16$ digraphs (Fig. 2) can be corresponded with its own item in:

$$
\begin{aligned}
x_1 B_1 = {} & a_{21}a_{31}a_{41}x_1^3 + (a_{21}a_{31}a_{42} + a_{21}a_{32}a_{41})\, x_1^2 x_2 \\
& + (a_{21}a_{31}a_{43} + a_{23}a_{31}a_{41})\, x_1^2 x_3 + (a_{21}a_{34}a_{41} \\
& + a_{24}a_{31}a_{41})\, x_1^2 x_4 + a_{21}a_{32}a_{42}x_1 x_2^2 + a_{23}a_{31}a_{43}x_1 x_3^2 \\
& + a_{24}a_{34}a_{41}x_1 x_4^2 + (a_{21}a_{32}a_{43} + a_{23}a_{31}a_{42})\, x_1 x_2 x_3 \\
& + (a_{21}a_{34}a_{42} + a_{24}a_{32}a_{41})\, x_1 x_2 x_4 + (a_{24}a_{31}a_{43} + a_{23}a_{34}a_{41})\, x_1 x_3 x_4 \,.
\end{aligned}
\tag{4.12}
$$

The items in this formula are presented in the same order as the corresponding to them digraphs presented in Fig. 2. The expression for $x_2 B_2$ can be derived from the formula (4.12) due to the replacement of the indexes 1 and 2 in subscripts in its items with 2 and 1, respectively. Similarly, one can obtain the expressions for $x_3 B_3$ or $x_4 B_4$ via the inversion of indexes 1 and 3 or 1 and 4 in formula (4.12). So by taking into account the relations (3.8) and (4.10) one can obtain the resultant relations of four-component copolymer compositions. There are no substantial difficulties in deriving similar formulae for the products of copolymerization of more than 4 types of monomers; however, the number of directed trees in every set becomes huge and thus to select all such digraphs directly will be quite tedious procedure without a computer. Nevertheless one can draw a number of general conclusions concerning the dependence of $\vec{\pi}$ on $\vec{x}$ and a_{ij} at arbitrary m.

Every component π_i of the composition vector $\vec{\pi}$ is the ratio of the uniform polynomials in variables x_1, x_2, ..., x_m of degree m. This statement follows from the fact that any directed tree with m points has just $(m - 1)$ arcs, and the end of each arc can be corresponded with a factor which is equal to the proper component of vector $\vec{x}$. In such a converging directed tree from any point, except the root, always comes out a single arc, and hence all the polynomial $x_1 B_1$ coefficients become: $a_{2\alpha} a_{3\beta} \dots a_{m\omega}$. Consequently, every ordered set of numbers $\{\alpha, \beta, \dots, \omega\}$ unambiguously characterizes the corresponding digraph. By means of a computer one can easily obtain the algorithm for selecting all such sets at given m and then put down an explicit formula for the instantaneous composition of the multicomponent copolymer, which is rather convenient for the calculation of the composition drift with the conversion (see Sect. 5).

4.4 Degenerated Systems

Several papers [116–118] theoretically considered a number of special cases of copolymerization of three monomers, some of which are incapable of forming homopolymers and/or certain binary copolymers. In such degenerated systems when some constants k_{ij} of propagation rates (2.8) are equal to zero, the explicit dependencies (3.8), (4.10), and (4.11) of the components of vectors $\vec{\pi}$ on $\vec{x}$ are the same as in the general case, i.e. the ratio of the polynomials of degree m. However, some of their coefficients may vanish and also contain in addition to $a_{ij} = k_{ij}/k_{ii}$ relative reactivity ratios $b_{ij}^s = k_{sj}/k_{si}$ of radicals of s-th type (corresponding to the monomer M_s incapable of homopolymerizing), under the propagation reactions of monomers M_j and M_i. Essentially, the kinetic parameters b_{ij}^s (in contrast to a_{ij}) can not be estimated via the analysis of binary copolymerization products, but only on the basis of the experimental data obtained in ternary systems. The expressions for terpolymer composition which is formed in some degenerated systems [47, 116–118], along with similar formulae for the degenerated systems, when the number of monomers m exceeds 3 [119], are easily derived [6] from the above general formulae (3.8) and (4.10) after their proper modification described below. At first, in expressions (4.10) one should put $a_{ij} = a_{ji} = 0$ for the couples of monomer M_i, M_j, which are incapable of copolymerizing. This corresponds to the exclusion from the digraph sets, shown in Figs. 1 and 2, of those trees in which points M_i and M_j are connected by an arc. Secondly, in formulae (4.10) for each monomer M_s which is incapable of homopolymerizing one should replace: (a) $a_{ss} = 1$ in σ_s with 0; (b) $a_{sj}(j = 1, \dots, m)$ with the s-th type radical reactivity ratios b_{ij}^s of the addition reactions of this radical with monomer M_j $(j = 1, \dots, m)$ in comparison with some reference monomer M_i.

A quite different type of degenerated system was considered in Ref. [120], the authors of which carried out a theoretical analysis of the run length distributions

$f^{(i)}(n)$ of the products of the terpolymerization of the peculiar monomer pairs, e.g., geometrical isomers like the esters of the maleic and fumaric acids. The peculiarity of such monomers consists in the fact that after their introduction into the macromolecule they become nondistinguishable and consequently, to both of them corresponds the same type of monomer unit. Hence the products of terpolymerization of such a pair of peculiar monomers with the third component (e.g. styrene) are binary, not triple, copolymers. One can easily demonstrate that the unit distribution of such copolymer even within the framework of the terminal model cannot be described by Markovian statistics. In particular, the conclusion made by the authors of Ref. [120] is that the expressions for $f^{(i)}(n)$ can assume a form different from the exponential Flory distribution. A comprehensive statistical description of the mentioned non-Markovian copolymers is carried out by means of the procedure of the conditional coloring of units of the peculiar monomers as it was done in Sect. 3.2 for the calculation of the copolymer statistical characteristics within the framework of the Seiner-Litt model. The distribution of the colored units in macromolecules is described by the Markov chain with three states S_1, S_2, S_3 which are connected with the states $\bar{M}_1$, $\bar{M}_2$ in sequences of non-colored units of the real binary copolymers in the following way $\bar{M}_1 = S_1, \bar{M}_2 = S_2 + S_3$. The probability of any one of such sequences is calculated through the same algorithm described in Sect. 3.2 but each vector $\pi^{(i)}(i = 1, 2)$ now has just i components, and each submatrix Q_{ij} consists of i rows and j columns. Using the above algorithm one can easily calculate any statistical characteristic of the non-Markovian copolymers under consideration.

4.5 Azeotropy

The instantaneous copolymer composition $\vec{X}$ generally doesn't coincide with the monomer feed composition $\vec{x}$ from which the copolymer was produced. Such a coincidence $\vec{X} = \vec{x}$ can occur only under some special values of monomer feed composition $\vec{x}$, called "azeotropic". According to definition these values can be calculated in the case of the terminal model (2.8) from a system of non-linear algebraic equations:

$$\pi_i(\vec{x}) = x_i \qquad (i = 1, 2, ..., m) \tag{4.13}$$

where the dependence of $\vec{\pi}$ on $\vec{x}$ is determined by the relationships (3.8) and (4.10). A great deal of literature is available dealing with the theoretical problems of azeotropy in the processes of copolymerization of three [121–124, 126–130, 135–142] or more monomers [46, 125, 131–134].

In the case of the terpolymerization one can reduce the non-linear set of equations (4.13) to a single equation of fourth degree [121–123], for which a rigorous explicit solution is obtained and the conditions of its existence are analyzed [121].

In addition to a true (triple) azeotrope, some authors suggested for m = 3 to consider "limited" (unitary) [123–125] and "partial" (binary) [126, 123] azeotropes. Both of them in contrast to the former one are located along certain azeotropic lines inside the Gibbs-Roozeboom triangle. Along the i-th unitary azeotropic line the fraction π_i of units $\bar{M}_i$ in the copolymer coincides with the fraction x_i of this monomer M_i in the reaction mixture. Along the binary azeotropic line for a given pair of monomers M_i and M_j the ratio π_i/π_j is equal to x_i/x_j. Three unitary azeotropic lines do correspond to each terpolymer system. If any two of them intersect inside the Gibbs-Roozeboom triangle, the third one should also pass through this intersection point, which is a true azeotrope. The intersection of the binary azeotropic lines is possible only at this very point; however, for some monomer pairs such lines may not exist at all [123].

The lines of the limited azeotropy according to O'Driscoll [124, 127] may be used for qualitative characteristics of the degree of composition inhomogeneity of the terpolymers of different compositions. With the same aim, Slocombe [128] put forward another graphical procedure according to which the compositions of monomer feed $\vec{x}$ and terpolymer $\vec{\pi}$ are joined by an arrow inside the Gibbs-Roozeboom triangle. The length of the arrow decreases when $\vec{x}$ approaches a true azeotrope $\vec{x}^*$, where all the arrows are contracted into a point. Both of the above graphical methods are used for the theoretical consideration of the real terpolymerization processes [122, 123, 125, 129]. The authors of Ref. [130] suggested an original procedure for the graphical representation of the copolymer composition dependencies on monomer feed composition, which, they believe, being the combination of O'Driscoll and Slocombe procedures, permits to eliminate their shortcomings. Considering multicomponent copolymerization theory, it is hard to overestimate the paper by Walling and Briggs [46] who managed to reduce the calculation of the azeotropic composition $\vec{x}^*$ in the system at any number of components m to the solution of the set of the linear equations:

$$x_i^* = \omega_i \sum_{j=1}^m a_{ij} x_j^*, \qquad \sum_{j=1}^m \omega_j a_{ji} = 1 \qquad (i = 1, 2, ..., m). \qquad (4.14)$$

Similar equations along with the various explicit forms of their solution were reported in a number of subsequent papers [99, 131–134], reviews [119, 135, 136], and in a monograph [6]. Note that having a physical meaning solutions $\vec{x}^*$ of both non-linear (4.13) and linear (4.14) sets of equations coincide as was conclusively proved in the general case at any m [46] and independently in the case of terpolymerization by virtue of another technique [137]. Consequently, when the Eq. (4.14) in the region $x_1 + x_2 + ... + x_m = 1$, called "m-simplex", have the positive solution $(x_1^* > 0, x_2^* > 0, ..., x_m^* > 0)$ it gives a rigorous value of azeotrope composition. Due to the linearity of Eq. (4.14) they have no more than one solution. The solution which has a physical meaning exists when the determinants D_i $(i = 1, 2, ..., m)$ of all matrices, which are obtained from the matrix of the reciprocal activities $\{a_{ij}\}$ by the substitution of row i elements for unity, are of the same sign. In this case the determinant D of matrix $\{a_{ij}\}$ has this sign too, and the values of $\omega_i = D_i/D$ at any $i = 1, 2, ..., m$ will be within the range $0 < \omega_i < 1$. Then knowing $\omega_1, \omega_2, ..., \omega_m$ one can calculate at once the azeotropic composition $\vec{x}^* = \vec{X}^*$ by performing the following modifications in (4.10) for $\vec{\pi}$. At first let us assume all $\sigma_i = 1$, then substitute x_i for ω_i and inverse indexes i and j in each factor a_{ij}. As a result, the expression for each component $\pi_i = \Delta_i/\Delta$ transforms into the expression for the component $x_i^* = X_i^* = \Delta_i^*/\Delta^*$. For example, in the case of terpolymerization, the application of the above algorithm to the formulae (4.11) leads to the following expressions [6, p. 270].

$$\Delta_1^* = \omega_1(a_{12}a_{13}\omega_1 + a_{12}a_{23}\omega_2 + a_{13}a_{32}\omega_3),$$

$$\Delta_2^* = \omega_2(a_{13}a_{21}\omega_1 + a_{21}a_{23}\omega_2 + a_{23}a_{31}\omega_3), \qquad (4.15)$$

$$\Delta_3^* = \omega_3(a_{12}a_{31}\omega_1 + a_{21}a_{32}\omega_2 + a_{31}a_{32}\omega_3),$$

which coincide within the accuracy of the insignificant common factor with the formulae obtained via the tedious calculations [121]. As it may be easily demonstrated, the parameters R, P, Q introduced in [121] respectively are equal to D_1, D_2, D_3 but have opposite sign. The dependencies of the latters on reactivity ratios can easily be obtained through formulae which are derived from the expression for determinant $D = |a_{ij}|$ at $m = 3$:

$$D = 1 - (a_{12}a_{21} + a_{13}a_{31} + a_{23}a_{32}) + a_{12}a_{23}a_{31} + a_{13}a_{32}a_{21} \qquad (4.16)$$

after the replacement of the elements of the proper row with unity

$$D_1 = 1 - (a_{21} + a_{31} + a_{23}a_{32}) + a_{23}a_{31} + a_{32}a_{21},$$

$$D_2 = 1 - (a_{12} + a_{32} + a_{13}a_{31}) + a_{12}a_{31} + a_{13}a_{32}, \qquad (4.17)$$

$$D_3 = 1 - (a_{13} + a_{23} + a_{12}a_{21}) + a_{12}a_{23} + a_{12}a_{21}.$$

Note that the expressions (4.15) can be obtained even without the formulae (4.11), if we employ an algorithm similar to the one used for the derivation of formulae (4.11). Actually, the proper diverging from the point i directed tree corresponds to each of the items in the expression for Δ_i^*. A weighting factor $\omega_i a_{ij}$ corresponds to each of the arcs in these digraphs that leaves point i and enters point j. The sum of all so weighted trees directed from a root of type i directly gives an expression similar to expressions (4.15) for the value of Δ_i^*, which is equal (when all ω_i are positive) within the accuracy of the normalizing factor Δ^* to the component $x_i^* = X_i^* = \Delta_i^*/\Delta^*$ of azeotropic composition under the copolymerization of an arbitrary number (m) of monomers.

As long ago as 1960, Tarasov et al. [121] presented some examples of the concrete three-component systems for which the existence of the azeotropic composition had already been predicted theoretically. The list of such systems was widened substantially after publication of the important paper [125], where a set of the known tabulated values of 653 pairs of reactivity ratios for a computer search of the possible multicomponent azeotropes was employed. For this aim one should, at first, reveal all the completely characterized multicomponent systems for which the values of reactivity ratios of all monomer pairs are tabulated. This problem can be formalized by reducing it to the search on the graph with 653 lines of a complete set of subgraphs with $m \geq 3$ points connected by the lines, the number of which is maximally possible at given m and is equal to $m(m - 1)/2$. In such a set among N_m ($m = 3, 4, \ldots$) completely characterized m-component systems a certain number N_m^* of systems having an azeotrope inside m-simplex were found [125]:

m	3	4	5	6	7	8	9
N_m	731	598	330	113	22	2	0
N_m^*	37	2	1	0	0	0	0

Among 37 terpolymer systems with inner (three-component) azeotrope 9, 19, and 9 have 3, 2, and 1 binary azeotropes, respectively, located on the sides of the Gibbs-Roozeboom triangle. Considering the systems in which such azeotropes are absent, no one containing a three-component azeotrope were found [125]. Later [138, 137] a possibility of the existence in principle of such systems was shown and also the necessary and sufficient conditions of their realization were formulated [6].

Hence, within the framework of the traditional kinetic model (2.8) there is a mathematically rigorous solution of the problem of the calculations of the azeotropic composition $\vec{x}^*$ under the copolymerization of any number of monomer types knowing their reactivity ratios, i.e. the elements of matrix $\{a_{ij}\}$. However, since the values of a_{ij} can be estimated from the experiment with certain errors δa_{ij}, the calculated location of azeotrope $\vec{x}^*$ is also determined with an accuracy, the degree of which is characterized by vector $\delta\vec{x}^*$ with components $\delta\vec{x}_k^*$ ($k = 1, 2, \ldots, m$) and modulus $|\delta\vec{x}^*|$

$$\delta x_k^* = \sum_{(ij)} \frac{\partial x_k^*}{\partial a_{ij}} \delta a_{ij}, \qquad T_{(ij)(rs)} = \sum_k \frac{\partial x_k^*}{\partial a_{ij}} \frac{\partial x_k^*}{\partial a_{rs}},$$
$$|\delta\vec{x}^*|^2 = \sum_k (\delta x_k^*)^2 = \sum_{(ij)} \sum_{(rs)} T_{(ij)(rs)} \delta a_{ij} \delta a_{rs}.$$
(4.18)

The square matrix $\mathbf{T}$ with elements $T_{(ij)(rs)}$ has $m(m - 1)$ rows (in accordance with the number of ordered pairs (ij) or parameters a_{ij}) and determines the parameter sensitivity of the azeotrope value towards the accuracy estimation of the reactivity ratios. Really, when their errors are the same, the deviation $|\delta\vec{x}^*|^2$ (4.18) of the theoretically predicted location of azeotrope will more or less depend on the values of elements of matrix $\mathbf{T}$. The calculation of such elements have no principal difficulties since an explicit dependence of $\mathbf{x}^*$ on parameters a_{ij} is known. In the case of the rather strong parameter sensitivity, when derivatives of x_k^* with the respect to a_{ij} are large, even comparatively small errors in δa_{ij} may result in substantial errors in calculations of $\vec{x}^*$ making it quite impossible to predict theoretically the existence or absence of an azeotrope in the given system. The examples of such systems were discussed earlier [125, 132, 135, 139] but as far as the author knows nobody has yet carried out the quantitative consideration of the parameter sensitivity by means of the expressions (4.18).

The determination of the regions, where the compositions of copolymer and monomer feed mixture do not coincide but have rather close values, is of the certain importance. Tarasov et al. [140] suggested for them a term "the regions of the approximate azeotropic composition" and pointed out the conditions of the existence of such regions in the case of the terpolymerization. Later they were called "near-azeotropic" [126] or "pseudoazeotropic" [127] regions and for their determination the graphical methods developed by O'Driscoll and Slocombe [123, 128] were applied under the analysis of the concrete systems. Zaitsev et al. [139, 141] for this aim using the procedure reported in [140] have described terpolymerization of styrene, methylmethacrylate and acrylonitrile. There is a viewpoint [126, 142] that the true azeotropes cannot be realized at all under the processes of multicomponent copolymerization but one can expect only the existence of the near-azeotropic domains. According to the author's point of view one can hardly agree with this very statement if we take into the account the above considerations.

4.6 Simplified Terminal Model of the Multicomponent Copolymerization

The mostly well-known discussion concerning multicomponent copolymerization theory is probably connected with so-called "simplified" equations put forward in papers [143–147] for the description of $\bar{\pi}(\bar{x})$ dependence. These equations can be obtained from the general Walling-Briggs equations (3.8), (4.10) via substituting in them expression a_{1i}/a_{i1} for B_i. The derivation of these simplified equations is based on the assumption that the rate of the monomer M_j addition to the radical R_i is equal to the rate of the monomer M_i addition to the radical R_j:

$$k_{ij}R_iM_j = k_{ji}R_jM_i \tag{4.19}$$

for all $m(m - 1)/2$ pairs of (ij). Let us remember that the general Eqs. (3.8) and (4.10) derived under the steady-state approximations $(dR_i/dt) = 0$ for any type i of radicals are obtained as a result of summing of both parts of formula (4.19) over all values of index j. While these approximations are well-known and valid for the kinetics of the radical reactions, the assumptions (4.19) are, generally speaking, arbitrary. In the case of terpolymerization, the necessary and sufficient condition of the validity of the above assumptions is the fulfillment of the following equality:

$$\Lambda^+ = \Lambda^-, \qquad \Lambda^+ = a_{12}a_{23}a_{31}, \qquad \Lambda^- = a_{13}a_{32}a_{21} \tag{4.20}$$

which according to Ham [126, 145–149] is regarded to be universal and hence may be effectively used for the determination of one of the six reactivity ratios knowing the remaining five. During the further discussion [150–156] concerning the universal character of equality (4.20), there were put forth numerous arguments both in favour and against it. For example, Mayo [150] using the tabulated values of the reactivity ratios has estimated the values of the parameter $H = \Lambda^-/\Lambda^+$ for 16 various terpolymers. He found that for many of these, one could observe considerable deviations of H from unity which contradicts the condition (4.20). According to O'Driscoll [151] these deviations are within the limits of accuracy of reactivity ratio measurements. The validity of this conclusion has been questioned by Tidwell and Mortimer [152] who carried out an accurate statistical analysis of errors. In accordance with the opinion of the authors of Refs. [153, 156], the question of the universal character of Equality (4.20) remains vague and requires a further statistical analysis along with more precise experimental data on reactivity ratios.

When the reactivity ratios r_{ij} can be expressed in terms of the parameters of the well-known "Q-e" scheme of Alfrey-Price [20, 157], the condition (4.20) always holds [147, 150] and in the case of terpolymerization the general Eqs. (3.8) and (4.10) transform into the simplified equation [158]. It is rather curious that similar equations have been derived at the end of the 1940s [159] within the framework of the Alfrey-Price scheme, being investigated even for the general case of copolymerization of arbitrary number m of monomer types.

The violation of the condition (4.20) for the particular system means that the latter can not be described by the "Q-e" scheme. However, in addition to this scheme, there are known some others, also applied for the description of the reactivity of the polymer radical in propagation reactions [160]. One such scheme proposed by Bamford [161, 162] was successfully used by Jenkins [163] for an interpretation of the experimental values [150] of parameter H for a number of ternary systems. For many of these the values of H are noticeably different from unity, as it has to be according to the predictions of the Alfrey-Price scheme, but are in satisfactory agreement with the values calculated through the Bamford scheme [163, 160].

4.7 Copolymer Symmetry

When the assumptions (4.19) are valid the Markov chain describing a sequence distribution in macromolecules is found to be reversible. It follows from the chain definition and relationships that:

$$P\{\bar{M}_i\bar{M}_j\} = P\{\bar{M}_j\bar{M}_i\} \qquad \pi_i\nu_{ij} = \pi_j\nu_{ji} \qquad (4.21)$$

which are obtained from Eqs. (4.19) parallel with the steady-state approximation for radical concentrations [147, 164]. Both formulae (4.21) are completely equivalent and are valid for any pairs (ij). Considering expressions (4.21) as the equations for the components of vector $\vec{\pi}$ and solving them one can obtain at once a formula for the multicomponent symmetrical copolymer composition. The symmetry property which consists in equality of probabilities $P\{U_k\} = P\{\hat{U}_k\}$ of any directed sequence U_k to its mirror symmetrical sequence $\hat{U}_k$, follows from the reversibility of the corresponding Markov chain [6]. When all reactions (2.8) rate constants are different from zero, the Markov chain theory leads to the following

$(m - 1) (m - 2)/2$ independent conditions of multicomponent copolymer symmetry [6]:

$$a_{1i}a_{ij}a_{j1} = a_{1j}a_{ji}a_{i1} \qquad (i = 2, 3, \ldots, m - 1; j = i + 1, \ldots, m) \quad (4.22)$$

which at $m = 3$ are simplified to the single equality (4.20). Though certain attempts were made [158, 165] to determine the number of independent conditions of symmetry at $m = 4$, the obtained results seem to be erroneous. As follows from the conditions (4.22), copolymer with $m \geq 4$ types of monomer units will be symmetrical when all terpolymers containing units $\bar{M}_1$ possess this property.

For the symmetrical copolymers the formula $x_i^* = \Delta_i^*/\Delta^*$ of the azeotropic composition has a surprisingly simple form since in this case $\Lambda_i^* = \omega_i/B_i = \omega_i a_{i1}/a_{1i}$, where the values of $\omega_i = D_i/D$ were determined above. Their expressions in a particular case of terpolymerization obtained by means of relations (4.16) and (4.17) lead to the well-known formulae [119, 166], where the azeotropic composition is expressed explicitly through the reactivity ratios r_{ij}. Note, that the terms "partial" and "limited" azeotropes were introduced initially for the processes of the production of symmetrical copolymers [126, 127].

4.8 Statistical Stationarity in Copolymer Description

In order to solve the problem concerning the statistical stationarity of a certain Markov chain which describes m-component copolymer, it is quite sufficient [6] to calculate the eigenvalues $\lambda_1, \lambda_2, \ldots, \lambda_m$ of matrix $\mathbf{Q}$ with the elements (4.1), i.e. to find the roots of its characteristic equation. The latter can be obtained by equating matrix $\lambda\mathbf{E} - \mathbf{Q}$ determinant to zero, where $\mathbf{E}$ denotes a unit matrix in which the units are located on its main diagonal and all other elements are equal to zero. Since the matrix $\mathbf{Q}$ is stochastic, the modulus of all its eigenvalues do not exceed unity and $\lambda_1 = 1$. The vast majority of the real multicomponent systems may be with sufficient accuracy described by the stationary Markov chains when the polymerization degrees exceeds ~ 10. One should exclude the special systems, which among the eigenvalues of matrix $\mathbf{Q}$ have in addition to $\lambda_1 = 1$ another eigenvalue with a modulus close to unity [6]. It is worth to mention that the closer these values to unity the greater the minimal number of units in copolymer molecules which is necessary for their statistical stationarity. For instance, in the case of the terpolymerization, the eigenvalues of matrix $\mathbf{Q}$:

$$\lambda_{2,3} = \tfrac{1}{2}\{T - 1 \pm [(T - 1)^2 - 4D]^{1/2}\} \quad (4.23)$$

are determined only by the values of its trace T and determinant D. The simple formula (4.23) allows one to determine conditions of the statistical stationarity without computer calculations, similar to ones performed earlier [167].

4.9 Calculations Within the Framework of Models Other than the Terminal one

All the data presented in this section were considered within the framework of the conventional copolymerization model (2.8). Let us now consider similar problems for systems described by other models. In the case of terpolymerization, Ham [47] has formulated the algorithm how using the induction method to go from the Alfrey-Goldfinger equations to the equations of the penultimate model and write them down for a particular case where the influence of the penultimate unit was pronounced only in one of the reactions (2.8). One can arrange the Markov chain of the second order with m^2 states which allows one to calculate any statistical characteristic of the multicomponent copolymer within the framework of the penultimate model. It should be noted that a generalization of the Seiner-Litt model (2.5) for copolymerization of m monomers which are capable of forming $m(m + 1)/2$ possible pair complexes, has no principal difficulties. Moreover, the principle of the conditional colouring of the units is still the same, and the sequence of the coloured units will form Markov chain of second order with 2m states [49, 60]. However, the practical importance of the application of the above models for calculating multicomponent systems is poor due to a rather great number of kinetic parameters, which, in contrast to the conventional model (2.8), can not be estimated from the experiments in the binary systems only.

At the end of the present section concerning the statistical description of copolymerization, one may conclude that up-to-now this traditional area of quantitative theory of macromolecular reaction could be regarded as complete. In the forthcoming section some other problems of copolymerization theory connected with its dynamics will be discussed. These problems being of a vital practical importance require specific approaches for their successful solution. The efficiency of such approaches will be demonstrated below considering concrete examples.

5 Copolymerization Dynamics

5.1 Statistical Characteristics of the Copolymers Produced at High Conversions

The expressions presented in Sects. 3 and 4 allow one to calculate any statistical characteristics of the copolymer produced at low conversions $p \ll 1$, when the drift of the monomer feed composition $\vec{x}$ with conversion may be neglected and hence we can put $\vec{x} = \vec{x}^0$ in all corresponding relations. In the reaction course on an increase of the conversion p the difference of $\vec{x}$ from its initial value $\vec{x}^0$ becomes, generally speaking, more pronounced resulting in the necessity of considering copolymerization dynamics.

In such a consideration the composition vector $\vec{x}$ determines the state of the reaction m-component mixture. This state according to the condition

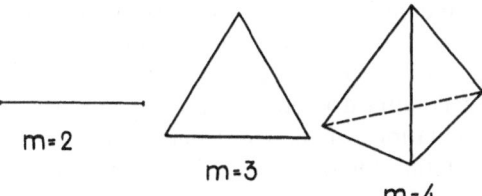

m=3

m=4

Fig. 3. Some m-simplexes for m-component copolymerization processes

$x_1 + x_2 + \ldots + x_m = 1$ is presented by a point in $(m - 1)$-dimensional phase space which is called "m-simplex" (see Fig. 3). Under the binary copolymerization, 2-simplex is a straight line segment with the unity length, since the distances x_1 and x_2 from any inner point to its end fit the condition $x_1 + x_2 = 1$. In the case of terpolymerization, 3-simplex $(x_1 + x_2 + x_3 = 1)$ is the well-known Gibbs-Roozeboom triangle, the sides of which are of unity lengths. In order to determine a composition which corresponds to some inner point inside this triangle, one should draw from it three straight lines, parallel to the sides of the triangle. The segments of these sides intersected by the crossing lines give the values x_1, x_2, x_3 of the composition under question. Each apex of the triangle corresponds to the homopolymerization of a certain monomer, and the side opposite to the apex — to the copolymerization of the rest two monomers. The phase space of the four-component copolymerization system is the inner part of a tetrahedron. Its faces, edges, and apexes correspond to all possible three-, two-, and single component systems, respectively, which can be produced from the set of the initial four monomers.

The evolution of the composition $\check{x}(p)$ in the course of the m-component copolymerization can be presented as the movement of the point, which characterizes the system state, inside m-simplex along some trajectory, the location of the point along which is determined by conversion p. The behavior of such trajectories is of vital importance since it determines the values of all statistical characteristics of the copolymers produced at high conversions.

The fractions of the arbitrary sequences $\{U_k\}$ are determined by the following formula:

$$\langle P\{U_k\}\rangle = \frac{1}{p} \int\limits_{0}^{p} P\{U_k\} \, dp' \qquad (5.1)$$

where the dependencies of the "instantaneous" values of the probabilities $P\{U_k\}$ on the monomer feed composition $\check{x}$ can be found by means of the standard procedures of the statistical copolymerization theory reported in Sects. 3 and 4. In order to calculate the sequence distribution by means of the formula (5.1) one should know a trajectory $\check{x}(p')$ of the system at all p' previous values of the conversion $0 \leqq p' < p$, since the integration in formula (5.1) is carried out just along this trajectory. This value is found by solving a set of differential equations:

$$(1 - p)\frac{dx_i}{dp} = x_i - X_i(\check{x}), \qquad x_i(0) = x_i^0 \qquad (i = 1, 2, \ldots, m) \qquad (5.2)$$

in which the dependence of the "instantaneous" copolymer composition $\vec{X}$ on the monomer feed composition $\vec{x}$ is obtained by considering the corresponding statistical problem within the framework of the particular kinetic model.

The operation of the conversional averaging defined by the angular brackets was proposed in Ref. [168] for the calculations of the run length distributions. This operation is also applied for calculating the other statistical characteristics of the copolymers produced at high conversions. For instance, in order to determine the composition distribution of such copolymers one should carry out the conversion averaging of the function (4.9). Since the latter is different from zero only within a very narrow composition range, it can be rather well approximated by the Dirac delta-function $\delta(\vec{\xi} - \vec{X})$. In this case the "instantaneous" component of the composition distribution is fully neglected in comparison with its main, i.e. "conversional", component, which is formed due to the drift of the monomer feed composition in the course of the synthesis. In practice it is quite sufficient to use the set of m one-dimensional partial distributions of each i-th component [6]:

$$\langle f_W^{(i)}(\xi_i)\rangle = \langle\delta(\xi_i - X_i)\rangle = \frac{1}{p}\sum_{p'}\left|\frac{dX_i}{dp}\right|^{(-1)} \tag{5.3}$$

where the sum is over those values of the conversion $p = p'$ at which $X_i(p)$ on trajectory inside m-simplex has the given value of ξ_i. In the case when $m \geq 3$ contrary to the binary copolymerization there may exist more than one such point, since in the multicomponent systems the dependencies $X_i(p)$, generally speaking, are not monotonic. In these cases, at conversions p^e where the function $X_i(p)$ reaches its extreme values, the amplitude of the corresponding distribution mode (5.3) becomes infinite. However, even when the X_i changes with conversion are of monotonic character, bimodal distributions (5.3) are possible. One of their maxima corresponds to the copolymer of the initial composition $\xi_i = X_i(0)$ and the other one to that of the final composition $\xi_i = X_i(1)$ (see Fig. 4).

The conditions of the existence of such bimodal distributions are well-known for the binary copolymerization [169–171], since in this case the differential equation proposed by Skeist [12] for the determination of the composition distribution has an explicit solution. Indeed when $m = 2$, the only one independent equation of the two equations (5.2) has a simple solution [169, 170, 172]:

$$1 - p = \left(\frac{x_1}{x_1^0}\right)^\alpha \left(\frac{1 - x_1}{1 - x_1^0}\right)^\beta \left(\frac{x_1 - x_1^*}{x_1^0 - x_1^*}\right)^\gamma \tag{5.4}$$

$$\alpha = \frac{r_2}{1 - r_2}, \quad \beta = \frac{r_1}{1 - r_1}, \quad \gamma = \frac{r_1 r_2 - 1}{(1 - r_1)(1 - r_2)};$$

$$x_1^* = \frac{1 - r_2}{2 - r_1 - r_2}$$

which determines within the framework of the terminal model the location of the reaction system inside 2-simplex with the conversion change p.

In Refs. [173–176] it was suggested to use the weight composition distributions instead of the molar ones and the results of their numerical calculation for some systems were reported. The authors of Ref. [177] carried out a thorough theoretical study of the composition distribution and derived an equation for it without the Skeist formula. They, as the authors of Ref. [178], proposed to use dispersion of the distribution (5.3) as a quantitative measure of the degree of the composition inhomogeneity of the binary copolymers and calculated its value for some systems. Elsewhere [179–185] for this purpose there were used other parameters of the composition distribution. In particular the discussion of the different theoretical aspects of the binary copolymerization is reported in a number of reviews by Soviet authors [186–189]. By means of numerical calculations there were analyzed [190–192] the limits of the validity of the traditional assumption which allows to ignore the "instantaneous" component of composition distribution of the copolymers produced at high conversions.

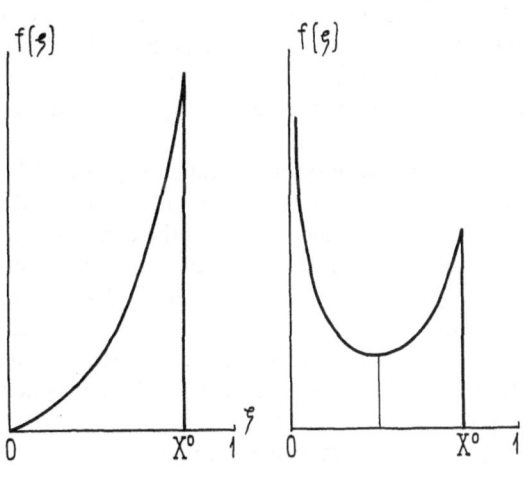

Fig. 4. Unimodal (1) and bimodal (2) composition distribution of binary copolymers prepared at complete conversion p = 1

The calculations of the statistical characteristics of such polymers within the framework of the kinetic models different from the terminal one do not present any difficulties at all. So in the case of the penultimate model, Harwood [193–194] worked out a special computer program for calculating the dependencies of the sequences probabilities on conversion. Within the framework of this model, Eq. (5.2) can be integrated in terms of the elementary functions as it was done earlier [177] in order to calculate copolymer composition distribution in the case of the simplified $(r_2 = r_2')$ penultimate model. In the framework of the latter the possibility of the existence of systems with two azeotropes was proved for the first time and the regions of the reactivity ratios of such systems [6] were determined. In a general version of the penultimate model (2.3–2.4) the azeotropic compositions $x_1^* = 1/(1 + \theta^*)$ are determined [6] by the positive roots $\theta = \theta^*$ of the following equation:

$$A\theta^3 + B\theta^2 + C\theta + D = 0,$$

$$A = -r_2'(1 - r_2), \qquad B = -1 + 2r_2' - 2r_1'r_2' + r_2r_1'r_2', \qquad (5.5)$$

$$C = 1 - 2r_1' + 2r_1'r_2' - r_1r_1'r_2', \qquad D = r_1'(1 - r_1).$$

The number of such roots is either even or odd, depending on the sign ($-$ or $+$ respectively) of the product $(1 - r_1)(1 - r_2)$. To carry out a further analysis one should calculate the sign of the following expression:

$$D = B^2C^2 - 4AC^3 - 4B^3D - 27A^2D^2 + 18ABCD. \qquad (5.6)$$

When $D < 0$, only one azeotrope exists in the system as in the case of the terminal model. When $D > 0$, the number $n = 0, 1, 2, 3$ of the azeotropic compositions coincides with the number of the sign changes in the sequence A, B, C, D of Eq. (5.5) coefficients. This procedure of finding the number n is equivalent to that proposed in Ref. [14] but is much more convenient for practical applications.

Under the copolymerization of more than two monomers Eqs. (5.3) cannot be integrated explicitly, and in order to determine the system trajectories one should need the numerical calculations. Examples of such calculations of the conversional change of composition and structure characteristics of the terpolymers have been reported in Refs. [195–200]. One should pay special attention to Ref. [200] where the programs for the computer realization of such calculations are presented. Under the copolymerization of four or more monomers, the composition drift with the conversion was calculated [7, 8] only within the framework of the simplified terminal model described above in Sect. 4.6.

As for the composition distributions (5.3) and their dispersions:

$$\sigma_i^2(p) = \frac{1}{p} \int_0^p X_i^2 \, dp' - \langle X_i \rangle^2, \qquad \langle X_i \rangle = \frac{x_i^0 - (1 - p) x_i}{p}. \qquad (5.7)$$

Examples of calculations of such characteristics for the concrete terpolymer are presented, for instance, in the monograph listed as Ref. [6]. Similar computer calculations of the trajectories $\vec{x}(p)$ and $\vec{X}(p)$ which determine the values of the statistical characteristics (5.3) and (5.7) can be in principle carried out for other processes of multicomponent copolymerization. However, the original approach [6, 201, 13–15, 18, 202, 203] allows one to use traditional methods of dynamic systems theory [204] to reveal the main qualitative peculiarities of the behavior of the trajectories $\vec{x}(p)$ and $\vec{X}(p)$ only via the elementary arithmetical operations by means of a pocket microcalculator.

5.2 The General Dynamic Theory of m-Component Copolymerization

The first important question we need to answer is how the monomer feed composition $\vec{x}(p)$ will be changed with conversion at various initial values $\vec{x}^0$ and the parameters of kinetic copolymerization model. When such a trajectory $\vec{x}(p)$ is known, on the base of the formulae (5.1), (5.3), and (5.7) one can find the main statistical copolymer characteristics at any number of its components within the framework of the chosen kinetic model.

Each trajectory $\vec{x}(p)$ is regarded to be a solution of the universal set of dynamic equations (5.2), the form of the right-hand parts of which is determined by the selected copolymerization model.

Any trajectory can end when $p \to 1$ at a stationary point (SP), in which all the right-hand parts of equations (5.2) equal zero. In the case of the terminal model (2.8) all such SPs are those solutions of the non-linear set of the algebraic equations (4.13) which have a physical meaning. Inside m-simplex one can find no more than one SP, the location of which is determined by the solution of the linear equations (4.14). In addition to such an inner azeotrope of the m-simplex, azeotropes can also exist on its boundaries which are n-simplexes ($2 \leqq n \leqq m - 1$). For each of these boundary azeotropes (m − n) components of vector $\vec{x}^*$ are equal to zero, so it is found to be an inner azeotrope in the system of the rest n monomers. Moreover, the equations (4.13) always have m solutions $x_i = \delta_{is}$ (where δ_{is} is the Cronecker Delta-symbol which is equal to 1 when $i = s$ and to 0 when $i \neq s$) corresponding to each of the homopolymers of the monomers M_s ($s = 1, ..., m$). Such solutions together with all azeotropes both inside m-simplex and on its boundaries form a complete set of SPs of the dynamic system (5.2).

Its qualitative analysis assumes establishing types of these points for determining the behavior of the trajectories $\vec{x}(p)$ in the vicinity of such SPs. In particular, this analysis permits to find all stable SPs. This is of practical importance since the trajectories $\vec{x}(p)$ under the condition of $p \to 1$ approach only these points. Each of them has its own basin of attraction which is defined as a set of all initial monomer feed compositions $\vec{x}^0$ for which the trajectories asymptotically approach the attractor (SP). Hence when $\vec{x}^0$ is located in the basin of a certain stable SP, the following drift of $\vec{x}(p)$ and $\vec{X}(p)$ is provided only to a given stationary point. Consequently, the whole m-simplex in accordance with the number of these points is divided into the same number of their basins. Each of them in its turn consists of one or more cells separated by the separatrix sets of m-2 dimension which have to be drawn inside the simplexes, in Fig. 3 by points, lines, and surfaces, respectively. Each of the above cells contains just one stable and one unstable SP. The whole trajectories inside the given cell reach a stable point and leave the unstable one [205]. These trajectories do not intersect with each other anywhere but at SP and completely cover all cell phase space. Hence, the trajectory which is part of the only whole trajectory passing through the given point $\vec{x}^0$ of the cell, unambiguously corresponds to a chemical system with a given initial monomer feed composition $\vec{x}^0$. The graphical representation of simplex showing the qualitative character of the mutual arrangement of the cells is qualified as its phase portrait [204, 205].

In the case of the binary copolymerization described by the terminal model (2.1) there are the following types of phase portraits:

$$1 \bullet\!\!\longrightarrow\!\!\circ 2 \text{ (I)} \qquad 1 \circ\!\!\longleftarrow\!\!\bullet\!\!\longrightarrow\!\!\circ 2 \text{ (II)}, \qquad (5.8)$$

$$1 \circ\!\!\longleftarrow\!\!\bullet 2 \text{ (III)} \qquad 1 \bullet\!\!\longrightarrow\!\!\circ\!\!\longleftarrow\!\!\bullet 2 \text{ (IV)} \qquad (5.9)$$

to which correspond values of the pairs of parameters (2.2):

$$r_1 > 1, \qquad r_2 < 1 \text{ (I)} \qquad r_1 < 1, \qquad r_2 < 1 \text{ (II)},$$

$$r_1 < 1, \qquad r_2 > 1 \text{ (III)} \qquad r_1 > 1, \qquad r_2 > 1 \text{ (IV)}. \qquad (5.10)$$

In the systems (I) and (III) 2-simplex consists of a sole cell, all the trajectories inside which approach SP corresponding to homopolymer M_s where $r_s < 1$. The systems (I) and (III) topologically are equivalent, since they differ from each other only by the inversion of the monomer indexes; therefore their phase portraits are of the same type, too. In the systems (II) and (IV) the azeotropic point separates the simplex into the two cells. However, the system (IV), in which both parameters r_1 and r_2 exceed unity practically is non-realizable [20–24]. That is why the stable binary azeotropes are excluded from the consideration, and the dynamics of the copolymerization of two monomers is exhaustively characterized by only two types (I) and (II) of phase portraits.

The above results in the case of the binary copolymerization can be easily obtained from the analysis of the expression (5.4). However, for the copolymerization of more than two monomers such an analysis is not possible since the proper explicit solution of Eqs. (5.2) is not available yet. The traditional methods of dynamic systems theory are considered the most effective for establishing the common qualitative peculiarities of the trajectory behavior [204–205].

Within such a consideration the type of any SP $\check{x}^*$ is determined by a set of roots of the corresponding characteristic equation:

$$\lambda^{m-1} + \alpha_1\lambda^{m-2} + \ldots + \alpha_{m-1} = 0, \qquad \alpha_k = (-1)^k S_k^{(m)} \qquad (5.11)$$

where each coefficient α_k represents a sum $S_k^{(m)}$ of all principle minors of order k of the matrix $E - C$ of order m within the accuracy of the sign. The elements of matrix C are determined by the following formulae

$$C_{ij} = \pi_{ij} + \frac{1}{m}, \qquad B_{ij}(\check{x}) \equiv \frac{\partial B_i}{\partial x_j},$$

$$\pi_{ij} \equiv \frac{\partial \pi_i}{\partial x_j}\bigg|_{\check{x}=\check{x}^*} = \omega_i a_{ij} - x_i^* \qquad (5.12)$$

$$+ \frac{1}{\Delta^*}\left[B_i^* \sigma_i^* \delta_{ij} - x_i^* B_j^* \sigma_j^* + x_i^* \left(\sigma_i^* B_{ij}^* - \sum_{s=1}^{m} x_s^* \sigma_s^* B_{sj}^* \right) \right]$$

where asterisk means that the corresponding function is calculated at the azeotropic point $\check{x} = \check{x}^*$.

In terms of the graph theory one can formulate an easy algorithm for finding through C_{ij} the coefficients α_k of the characteristic equation (5.11). The procedure is rather simple. One should consider the diverging directed trees which are obtained from the converging directed trees used in Section 4.3 for the calculation of $\check{\pi}(\check{x})$ replacing the directions of all arcs with the opposite ones. The set of the trees in the graph theory is called "forest" [115]. In future we shall use the term "(m, k)-forest" to denote the graph (generally speaking, unconnected) with m points and k arcs, all the connected components of which are regarded to be the trees. Let us assume that each arc $\{ij\}$ has a certain weight C_{ij}, and the weight of the forest is proposed to be the product of the weights of all its arcs. Then the factor $S_k^{(m)}$ (5.11) which determines the value of the coefficient α_k is found to be equal to the total weight of all possible (m, k)-forests consisting of diverging directed trees.

The calculation of coefficients α_k of equation (5.11) for SP located on m-simplex boundary is reduced to the analysis of the inner azeotropes in the systems with the less number of component. Let us consider, for example, one of such boundary points $\check{x}^*$ where $x_1^* > 0$, $x_2^* > 0$, ..., $x_l^* > 0$, $x_{l+1}^* = 0$, ..., $x_m^* = 0$. This SP obviously is the inner azeotrope on l-subsimplex (12 ... l), where its location and corresponding characteristic equation of $(l - 1)$ degree are determined as a result of the consideration of copolymerization of following monomers M_1, M_2, ... M_l. The obtained solutions λ_1, λ_2, ... λ_{l-1} of the above equation are also the roots of the characteristic equation (5.11) of m-component system, and the rest $(m - l)$ roots are calculated by the following formula

$$\lambda_{i-1} = D_i(12 \dots li)/D(12 \dots l) \qquad (i = l + 1, \dots, m). \qquad (5.13)$$

D_i and D symbols for the proper determinants have been already introduced in the Sect. (4.5), and the figures in the brackets denote the monomer set in the system, for which these determinants are calculated.

For the boundary SP $x_i^* = \delta_{i1}$ which is located at the point (1) of m-simplex, corresponding to the homopolymer of monomer M_1, formula (5.13) yields

$$\lambda_{i-1} = D_i(i1)/D(1) = 1 - a_{1i} \qquad (i = 2, 3, \dots, m). \qquad (5.14)$$

When the boundary azeotrope is located on m-simplex edge (12), corresponding to the binary copolymer of monomers M_1 and M_2 the roots of the characteristic equation (5.11) are equal to

$$\lambda_1 = \frac{-D_1(12) D_2(12)}{D(12)} = \frac{(a_{12} - 1)(a_{21} - 1)}{a_{12}a_{21} - 1} \qquad (3 \leq i \leq m),$$

$$(5.15)$$

$$\lambda_{i-1} = \frac{D_i(12i)}{D(12)} = \frac{(a_{12} - a_{1i})(a_{21} - a_{2i}) - (a_{1i} - 1)(a_{2i} - 1)}{a_{12}a_{21} - 1}.$$

The stationary point of the system (5.2) is by a definition stable one, if all the roots of its characteristic equation (5.11) have the negative real parts. The Routh-Hurwitz criteria presented in Ref. [206] permits escaping the calculations of these roots to establish the simple relations between the coefficients α_k, which allow to point out simple stability conditions. For instance, in the case of terpolymerization the positivity of both coefficients α_1 and α_2 is regarded to be a criteria of such stability, and as for four-component copolymerization the following non-equality $\alpha_3 < \alpha_1 < \alpha_2$ has also to be hold. At arbitrary number m of the components the positivity of all α_k is regarded to be necessary (but not sufficient) stability condition. For the stability of the boundary SP of m-component system located inside the certain boundary l-subsimplex of monomers M_1, M_2, ..., M_l the stability of the above SP in such subsimplex and negativity of all values of λ_l, λ_{l+1}, ..., λ_{m-1} (5.13) are needed.

For the stability of SP $x_i^* = \delta_{is}$ located at s-th apex of m-simplex it is necessarily and sufficiently for non-equality $r_{si} < 1$ to be hold at all $i \neq s$. It was found that even the availability of one of such stable SP in the system excludes the possibility of the existence there of the stable inner azeotrope. This important property ensues directly from the comparison of the expression for coefficient α_1 of Eq. (5.11) for the inner azeotrope:

$$\alpha_1 = \sum_{i=1}^m \omega_i - 1 - \sum_{i=1}^m \sum_{j=1}^m (d_{ij}^* - d_{ii}^*) x_i^* x_j^*, \qquad d_{ij}(\check{x}) = \frac{\sigma_i B_{ij}}{\Delta} \quad (5.16)$$

with the second formula (4.14). Consequently, in practice the inner azeotrope stability is to be checked up only in the systems where the stable SPs corresponding to the homopolymers are absent and, moreover, the sum of all ω_i exceeds unity. The expression (5.12) for the inner azeotropes can be reduced to:

$$\pi_{ij} = \omega_i a_{ij} - 2x_i^* + \delta_{ij} + x_i^* \left(d_{ij}^* - \sum_{s=1}^m x_s^* d_{sj}^* \right). \qquad (5.17)$$

One can propose [202, 203] the following general classification of the multicomponent copolymerization systems from the viewpoint of their dynamic behavior. This classification involves three subsequent hierarchical levels of detailing. At first, knowing only the signs of $(1 - a_{ij})$ values one can easily establish the directions of the trajectory movements along all edges (ij) on the m-simplex which afterwards can be corresponded to a certain digraph involving the points having degrees $(m - 1)$ and 2. The former points correspond to the apexes of the m-simplex and the latter ones to the binary azeotropes on its edges. All polymer systems resulting in the same (isomorphic) digraphs are believed to be of the same type. Inside it some subtypes may exist which can not be transformed into each other by turns in $(m - 1)$-dimensional space. For example, the systems (I) and (III) of the binary copolymerization (5.9) are regarded to be such subtypes. For the sake of simplicity we shall not distinguish them below at all.

Systems of the same type can differ by their genus in accordance with the values of the topological Poincare indexes of all their SPs. The value of the above index

Ind($\mathbf{x}^*$) of such SP $\check{\mathbf{x}}^*$ is equal to $(-1)^q$, where q is the number of the negative roots of its characteristic equation (5.11). Obviously, the sign of Ind($\mathbf{x}^*$) coincides with the sign of the expression $(-1)^{m-1}\alpha_{m-1}$.

The copolymerization of m-component systems can be classified by their genera in accordance with the numbers of the topologically different SPs inside the m-simplex and on all its boundary n-subsimplexes (n = 1, 2, ..., m − 1). Let us denote through N_n^+ and N_n^- the numbers of SPs in the system n-subsimplexes having the Poincare indexes +1 and −1, respectively. Then the set of such pairs of numbers $\{N_n^+, N_n^-\}$ at all n = 1, 2, ..., m will characterize the system genus. So it was discovered that not any of such sets may be realized in practice but only those that meet the following condition [202, 203]:

$$\sum_{n=1}^{m} 2^n \, (N_n^+ - N_n^-) = 1 + (-1)^{m-1}. \qquad (5.18)$$

Turning over all the sets meeting "the rule of azeotropy" (5.18) with the account for $N_1^+ + N_1^- \equiv N_1 = m$ and condition $N_n^+ + N_n^- \equiv N_n \leq m!/n!(m-n)!$ ($2 \leq n \leq m$) excluding the existence of more than one azeotrope at any n-subsimplex, one can easily find all possible genera − from the topological viewpoint − of m-component systems. However, not all of them are inevitably realized in practice since some genera even being allowed topologically can be forbidden within the framework of the above copolymerization model. The simplest example is the binary system (IV) (5.9) where $N_1^+ = 2, N_1^- = 0; N_2^+ = 0, N_2^- = 1$.

The most thorough topological classification of the dynamic behavior of the copolymerization systems suppose to make them out by their kinds, each of them is determined by the types of all SPs as well as by manifolds separating their basins (regions of their attraction). Every kind is characterized by the type of its phase portrait. The case of the binary copolymerization obviously is trivial since each of the types (5.8) consists of only one genus which in its turn includes a single kind.

In the case of degenerated systems (see Sect. 4.4), where monomers incapable of homopolymerization are present, they could remain unpolymerized after the end of the process. Each of these monomers could be assigned to a stable stationary point (SP) located at the s-th apex of the m-simplex. If the initial monomer feed composition $\check{\mathbf{x}}^0$ lies within the basin of attraction of such a SP, the copolymerization process is completed at conversion $p^{**} < 1$ when all monomers, except M_s, have already reacted. Its residual concentration M_s^{**} could be rather simply expressed through p^{**}, the value of which for binary copolymerization is determined by a trivial expression [168, 6, p. 239] and for $m \geq 3$ can be estimated from a numerical solution of a set of Eqs. (5.2). The dependence of $\vec{\mathbf{X}}$ on $\check{\mathbf{x}}$ in their right-hand parts could be easily determined taking into account the algorithm formulated in Sect. 4.4. It is worth mentioning that such degenerated systems are consistent with the above-mentioned general classification.

The above classification may be considered to be exhaustive for the most wide-spread simple (from the viewpoint of their dynamics) multicomponent systems where all of the whole trajectories $\check{\mathbf{x}}(p)$ and $\vec{\mathbf{X}}(p)$ start and finish at unstable and stable SPs, respectively. However, there are some other types of dynamic system attractors, for instance, limited cycles leading to the auto-oscillating regimes when the compositions both of monomer feed mixture and copolymer are oscillating with conversion. The features of such regimes will be discussed below for the case of three-component systems, for which an application of the above general approach will be demonstrated in detail.

5.3 Terpolymerization

In this case, in accordance with Fig. 3, the Gibbs-Roozeboom triangle is regarded to be the phase space, where one can find an attractor belonging to some type among those presented in Fig. 5. The attractor type can easily be determined from Table 5.1 by the signs of the SP characteristic equation (5.11) coefficients α_1, α_2 and by the value of $\beta = \alpha_1^2 - 4\alpha_2$. The above coefficients can be calculated rather simply by means of formulae (5.12) and (5.17) for any concrete system using the values of its parameters $r_{ij} \equiv 1/a_{ij}$.

However, it should be noted that even these elementary calculations are not needed, since the absence or presence of the inner azeotrope can be rigorously

Fig. 5. Possible types of attractors inside 3-simplex: node (*1, 2*), focus (*3, 4*), saddle (*5*) and limit cycle (*6, 7*). Stable (*1, 3, 6*) and unstable (*2, 4, 5, 7*) attractors

Table 5.1 Types of inner azeotropes in three-component copolymerization

N	α_1	α_2	β	Type of SP x* (x_1^*, x_2^*, x_3^*)	Ind(x*)
1	+	+	+	Stable node (SN)	+1
2	+	+	−	Stable focus (SF)	+1
3	−	+	+	Unstable node (UN)	+1
4	−	+	−	Unstable focus (UF)	+1
5		−		Saddle (S)	−1

Table 5.2 Types of boundary SPs in three-component copolymerization

λ	Type of SP x^* $(0, x_2^*, x_3^*)$	Ind(x^*)	q	Type of SP x^* $(0, 0, 1)$	Ind(x^*)
$-$	Saddle	-1	0	Unstable node	$+1$
$+$	Unstable node	$+1$	1	Saddle	-1
			2	Stable node	$+1$

established through "the rule of azeotropy" (5.18) knowing only the types of all SPs on the triangle boundary. It follows from Table 5.2 that the type of the binary azeotrope at the side (23) of the triangle (123) is unambiguously given by the sign of the root $\lambda = D(123)/D(23)$ (5.15) of the characteristic equation since its other root (5.15) due to $a_{23} > 1, a_{32} > 1$ is always positive. The number of the negative roots of Eq. (5.11) for any SP located at the triangle apex completely determines, in accordance with Table 5.2, the type of such a point. Moreover, the value of q for the j-th apex is equal to the number of the parameters a_{ji} exceeding 1.

When the types of all boundary SPs and their topological indexes are established by means of Table 5.2, one should use the "rule of azeotropy" (5.18). In the case of terpolymerization this rule can be written as:

$$\delta \equiv N_3^+ - N_3^- = (N_2 - N_1^+)/2 - N_2^+ + 1 \tag{5.19}$$

where N_1^+ and N_2^+ denote the number of SPs with index $+1$ (i.e. nodes) at the apexes and the sides of the triangle, respectively, and N_2 is the overall number of the binary azeotropes on its sides. To an arbitrary three-component system with given triple numbers $[N_1^+, N_2, N_2^+]$ corresponds, according to (5.19), a certain value of the topological parameter δ. It may be equal to only one of three numbers $+1, 0, -1$, to each of which according to Table 5.1 inside the triangle corresponds node or focus, the absence of azeotrope, and saddle, respectively.

In accordance with the general principles of the above classification the three-component systems can be divided into 7 types which are shown in Fig. 6. Each of the diagrams I, II, IV, VII may have two subtypes. The type of the concrete system which is characterized by a set of arrow directions on the triangle sides depends only upon a matrix of signs of values $1 - a_{ij}$.

In order to classify the systems in terms of their genus for each system one should also: (i) establish the types of the binary azeotropes by determining the signs of the proper values of λ (see Table 5.2) and (ii) estimate the topological parameter δ from formula (5.19). According to Table 5.2 the separatrix of the binary azeotrope enters it when $\lambda < 0$ and leaves it when $\lambda > 0$, and the number

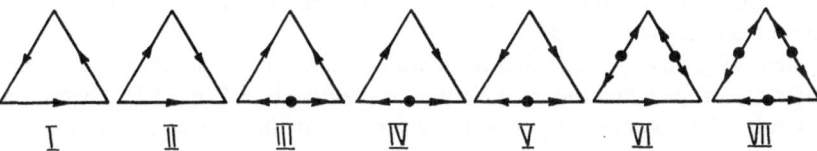

Fig. 6. Types of three-component systems

Table 5.3 Classification of three-component systems without limit cycles

Type	Genus	N_1^+	N_2	N_2^+	δ	Kind	N^0 [6]
I	1	0	0	0	+1	1, 2	1, 17
II	2	1	0	0	0	3	2
III	–	1	1	0	+1	–	3
III	3	1	1	1	0	4	5
IV	4	1	1	0	+1	5	4
IV	5	1	1	1	0	6	6
V	6	3	1	0	0	7	7
V	7	3	1	1	−1	8	8
VI	8	2	2	0	+1	9	9
VI	9	2	2	1	0	10, 11	10, 11
VI	10	2	2	2	−1	12	12
VII	11	3	3	0	+1	13	13
VII	12	3	3	1	0	14	14
VII	–	3	3	2	−1	–	15
VII	–	3	3	2	−1	–	16

of azeotropes of the latter type in the system is just equal to N_2^+. Varying N_2^+ within an interval from 0 to N_2 for each of the system types shown in Fig. 6 with the given number values N_1^+ and N_2 we can get the complete set consisting of 16 "pretenders", passing through the first step of the selection. As a second step we should reject the "pretenders" which do not fit the topological restrictions, since parameter δ is found to be different from one of the three meaningful values -1, 0, $+1$. Having rejected only one such system [3, 3, 3] for which $\delta = 2$ we obtain a list consisting of 15 "pretenders", presented in Table 5.3, corresponding to the reported classification of the three-component systems [6, p. 273; 13]. A more detailed analysis carried out afterwards [202, 203] revealed that three of these 15 "pretenders" are principally unrealizable within the framework of the above kinetic model, so only 12 different genera are left finally, the complete list of which is presented in Table 5.3.

It follows from Table 5.3 that the systems of the six genera have no inner azeotrope ($\delta = 0$), and for two other ones, namely 7 and 10, this azeotrope is regarded to be a saddle point ($\delta = -1$). For the other four systems which are characterized by $\delta = +1$ the azeotrope may be, according to Table 5.1, either the node (when $\beta > 0$) or the focus (when $\beta < 0$). Moreover, in the systems with the following genus numbers 4, 8, 11 the inner azeotrope is always unstable since they have stable SPs at the triangle apexes. Hence each of these genera has two kinds ($\beta > 0$ and $\beta < 0$), and the other ones listed in Table 5.3 excluding the genera 1 and 9 have only single kind. The system of genus 9 has two kinds differing from each other only by the direction of the separatrix connecting the binary azeotropes at the triangle edges.

The system of genus 1 is regarded to be an exclusive one since it only has the stable inner azeotrope when the coefficient α_1 is positive (see Table 5.1). Therefore the four kinds of the system of genus 1 correspond to the four combinations of signs of the α_1 and β values. If, as it is usually admitted under the topological classification of the dynamic systems, one doesn't distinguish for the sake of

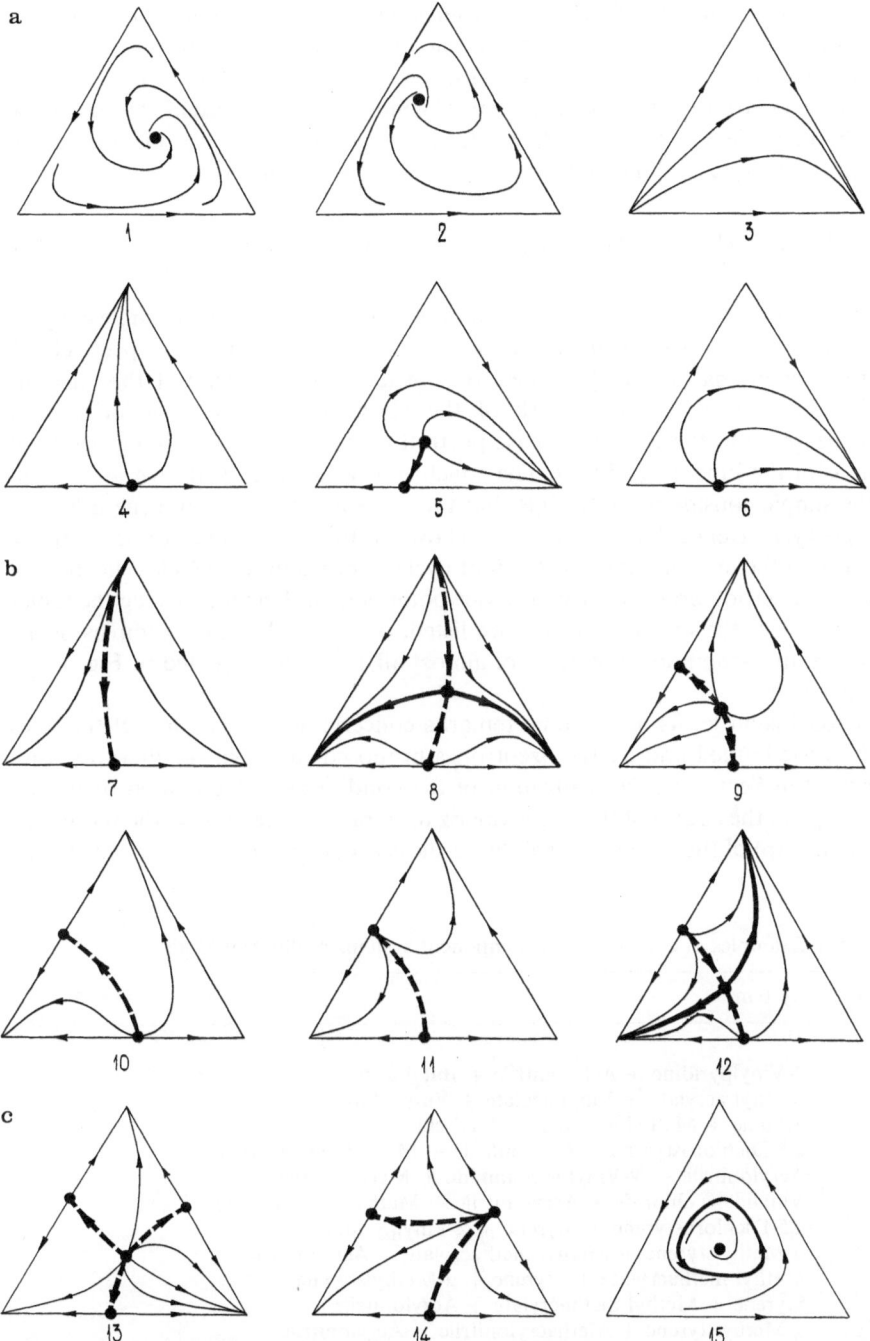

Fig. 7a–c. Possible kinds of phase portraits of terpolymerization processes. Separatrix, which separate cells in contrast to the trivial trajectories, are depicted by *heavy lines*. (These lines are *dotted* or *solid* depending on whether they are boundaries of the basins of attraction of SPs or not)

simplicity the nodes and foci, then any terpolymerization system, which contains no limit cycles, has a phase portrait of one of 14 kinds shown in Fig. 7.

The consideration of the systems with limit cycles reveals additional kinds of the phase portraits. Thus for the system of genus 1 there has been formulated [14, 15, 18] a sufficient condition $\alpha_1 \Lambda < 0$ for the existence of at least one limit cycle, which is stable ($\alpha_1 < 0$) or unstable ($\alpha_1 > 0$). The parameter:

$$\Lambda = (1 - a_{12})(1 - a_{23})(1 - a_{31}) + (1 - a_{13})(1 - a_{32})(1 - a_{21}) \qquad (5.20)$$

characterizing the behavior of the trajectories in the vicinity of the triangle contour is positive or negative when this contour is repeller or attractor, respectively. A more detailed analysis [203] revealed that in the system of genus 1 the unstable limit cycle within the framework of the above kinetic model is principally unrealizable, and therefore the phase portrait of this system containing one limit cycle is of the kind 15 in Fig. 7. The conclusion which can be drawn as a result of the simple consideration [203] is that there are no other system genera having a single cycle. Hence the set of 15 kinds shown in Fig. 7 is believed to be complete (correct to the even number of the limit cycles) according to the classification of terpolymerization processes from the viewpoint of their dynamics. Using the tables [20–24] where the reactivity ratios are listed, one can select the examples of the concrete three-component systems of almost all 15 kinds presented in Fig. 7 (see Table 5.4).

In conclusion let us make some remarks concerning the existence of the lines both of the limited and partial azeotropes proposed earlier in the literature and discussed in Sect. 4.5. The realization of the conditions of the limited or partial azeotropy in the course of the terpolymerization process means that the representing point $\check{x}(p)$ of the system doesn't leave these azeotropic lines under the change

Table 5.4 Examples of concrete three-component systems of different kinds

Kind	System
1	
2	2-Vinylpyridine + Acrylonitrile + Butyl acrylate
3	Methyl acrylate + Vinyl acetate + Vinyl chloride
4	Styrene + Methyl acrylate + Vinyl chloride
5	2,5-Dichlorostyrene + Acrylonitrile + Methyl methacrylate
6	Acrylonitrile + N-Vinyl succinimide + Methyl acrylate
7	Vinylidene chloride + Acrylonitrile + Methyl methacrylate
8	2,5-Dichlorostyrene + Styrene + 2-Vinylpyridine
9	α-Methylstyrene + Methyl methacrylate + Acrylonitrile
10	Methyl methacrylate + Styrene + α-Methylstyrene
11	Styrene + Methyl methacrylate + Acrylonitrile
12	α-Methylstyrene + Methacrylonitrile + Acrylonitrile
13	Styrene + Methyl methacrylate + 4-Vinylpyridine
14	Methyl methacrylate + Methacrylonitrile + Styrene
15	

of p. However, since these lines are not trajectories, the movement of this point along them is impossible. This simple forcible arguments let us draw the obvious conclusion that the famous discussion concerning the existence of limited and partial azeotrops seems to be over.

5.4 Copolymerization of Four Monomers

In this case according to Fig. 3 a phase space is found to be a tetrahedron, inside which an azeotrope of one of the eight types, presented in Fig. 8, may be located. In order to know which of them is to be realized one should find the signs of coefficients $\alpha_1, \alpha_2, \alpha_3$ of characteristic equation (5.11) and also their combinations:

$$\alpha = \alpha_1\alpha_2 - \alpha_3,$$

$$\beta = \alpha_1^2\alpha_2^2 - 4\alpha_1^3\alpha_3 - 4\alpha_2^3 + 18\alpha_1\alpha_2\alpha_3 - 27\alpha_3^2.$$

$$(5.21)$$

When $\beta > 0$, using Table 5.5 one can immediately point out the type of the inner azeotrope, and when $\beta < 0$, only the node (N) and saddle (S) in this table are replaced with the focus (F) and saddle-focus (SF), respectively. The sign of Ind $(\tilde{x}^*)$ of this azeotrope is opposite to the sign of α_3 and is determined by indexes of the boundary SPs (see Table 5.6) in accordance with the "rule of azeotropy" (5.18) which in the case of tetrapolymerization yields:

$$\delta \equiv N_4^+ - N_4^- = (2 + N_2 + 2N_3 - N_1^+ - 2N_2^+)/4 - N_3^+. \quad (5.22)$$

The types of SPs located at the apexes and edges of the tetrahedron can be established at once according to the right part of Table 5.6 without any additional calculations. For the first of the above SPs in each j-th point it is sufficient to indicate a number q of the parameters a_{ij} exceeding unity. The type of the four-component azeotrope, located at the edge (kl) of the tetrahedron is unambiguously characterized by the types of the two azeotropes, which correspond to it in three-component systems (ikl) and (jkl) represented by the tetrahedron faces

Table 5.5 Types of inner azeotropes in four-component copolymerization when $\beta > 0$

α_1	α_2	α_3	α	Type of SP x^*	Ind(x^*)
+	+	+	+	Stable node	-1
+	+	+	−	2nd order saddle	-1
−	+	+		– – – –	-1
+	−	+		– – – –	-1
−	−	+		– – – –	-1
+	+	−		1st order saddle	$+1$
+	−	−		– – – –	$+1$
−	+	−	+	– – – –	$+1$
−	+	−	−	Unstable node	$+1$
−	−	−		1st order saddle	$+1$

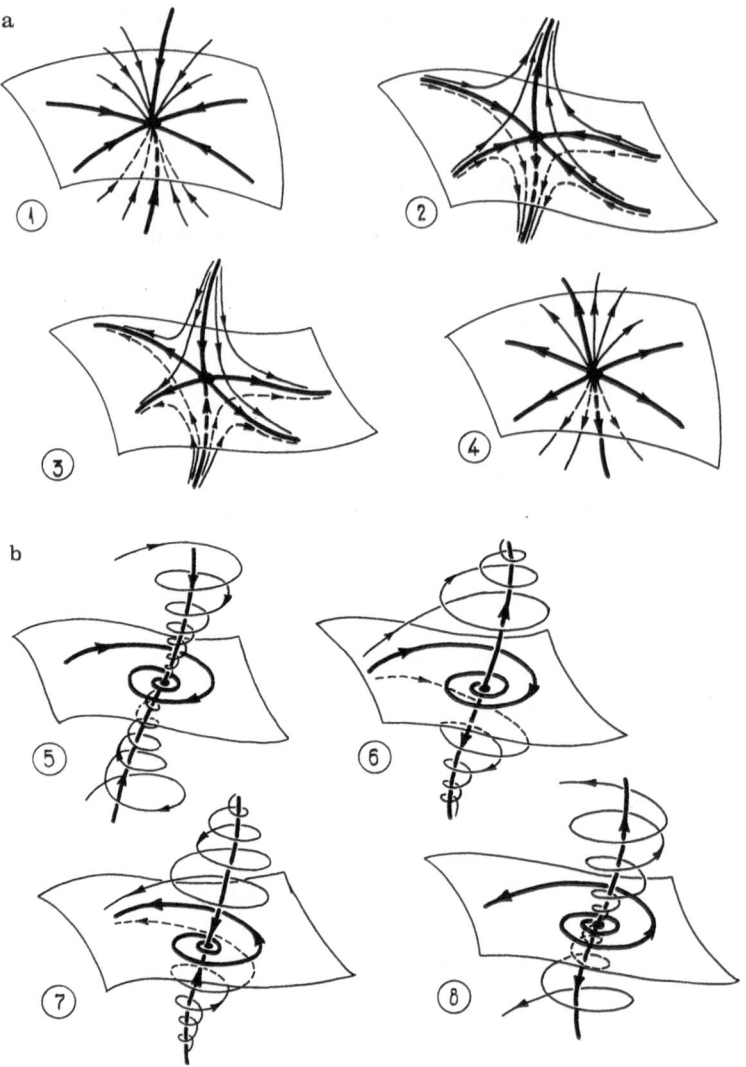

Fig. 8a, b. Possible types of SPs inside 4-simplex: stable (*1*) and unstable (*4*) node, saddle of the first (*2*) and second (*3*) order, stable (*5*) and unstable (*8*) focus, saddle-focus of the first (*6*) and second (*7*) order

intersecting at the edge (kl). Let us note, that through the saddle of the first order (S1) the separatrix surface will pass for certain. On the separatrix surface all the trajectories in the vicinity of the azeotrope approach this saddle, while out of this surface they will never reach it. Note that through the saddle of the second order (S2) such a surface inside the tetrahedron can not pass. The azeotrope type at the face (jkl) of the tetrahedron (ijkl) is determined according to the left part of Table 5.6 not only by the type of the azeotrope corresponding to it in the system consisting of three monomers M_j, M_k, M_l but also by the sign of the value of

Table 5.6 Types of boundary SPs in four-component copolymerization

λ	SP type in the system:		Ind(x*)	SP type in the system:			Ind(x*)
	(jkl)	(ijkl)		(ikl)	(jkl)	(ijkl)	
−	SN	SN	−1	UN	UN	UN	+1
+	SN	S1	+1	UN	S	S2	−1
−	SF	SF	−1	S	UN	S2	−1
+	SF	SF1	+1	S	S	S1	+1
−	UN	S2	−1				
+	UN	UN	+1	q	Type of SP x* (0, 0, 0, 1)		Ind(x*)
−	UF	SF2	−1	0	UN		+1
+	UF	UF	+1	1	S2		−1
−	S	S1	+1	2	S1		+1
+	S	S2	−1	3	SN		−1

$\lambda = D_i(jkli)/D(jkl)$ (5.13). The sign of the latter is exactly the same as $Ind(\overset{*}{x}{}^*)$ when the three-component azeotrope in the system (jkl) is the node or focus, and its sign is opposite to the sign of $Ind(\overset{*}{x}{}^*)$ when the above azeotrope is the saddle. The separatrix surface inside the tetrahedron surely passes through the azeotrope located on its face when in the ternary system corresponding to this face such an azeotrope is the saddle.

Using the expression (5.22) together with Tables 5.5 and 5.6 on the base of the general principles reported in Sect. 5.2 one can carry out an exhaustive classification of the four-component systems as it has been already done for terpolymerization in Sect. 5.3. However, when the forth monomer is added, the number of the system types increases from 7 (see Fig. 6) to 41 (see Fig. 9) and that is a reason why the results of the complete theoretical analysis cannot be represented in the framework of this review. Without appealing to the classification and using only the algorithm described in Sect. 5.2 one may present a phase portrait of any concrete four-component system and hence predict the qualitative character of its dynamic behavior before the computer calculations of trajectories $\overset{*}{x}(p)$ are performed.

Let us consider, for example, the copolymerization of styrene M_1, methylmethacrylate M_2, 4-vinylpyridine M_3, and methylacrylate M_4, for which by using the table reported in Ref. [21] one can write down the following matrix of reactivity ratios:

$$\begin{bmatrix} 1 & 0.52 & 0.54 & 0.75 \\ 0.46 & 1 & 0.57 & 1.5 \\ 0.70 & 0.79 & 1 & 1.7 \\ 0.18 & 0.3 & 0.22 & 1 \end{bmatrix}. \tag{5.23}$$

It follows from this matrix that this system belongs to type XXXVIII presented in Fig. 9. In order to represent its phase portrait one has to determine the type of all SPs, the specific lines, and surfaces. Three SPs located at the tetrahedron apexes and corresponding to the first three monomers are stable nodes and the forth one is the saddle of 2-nd order. The elementary analysis allows one to

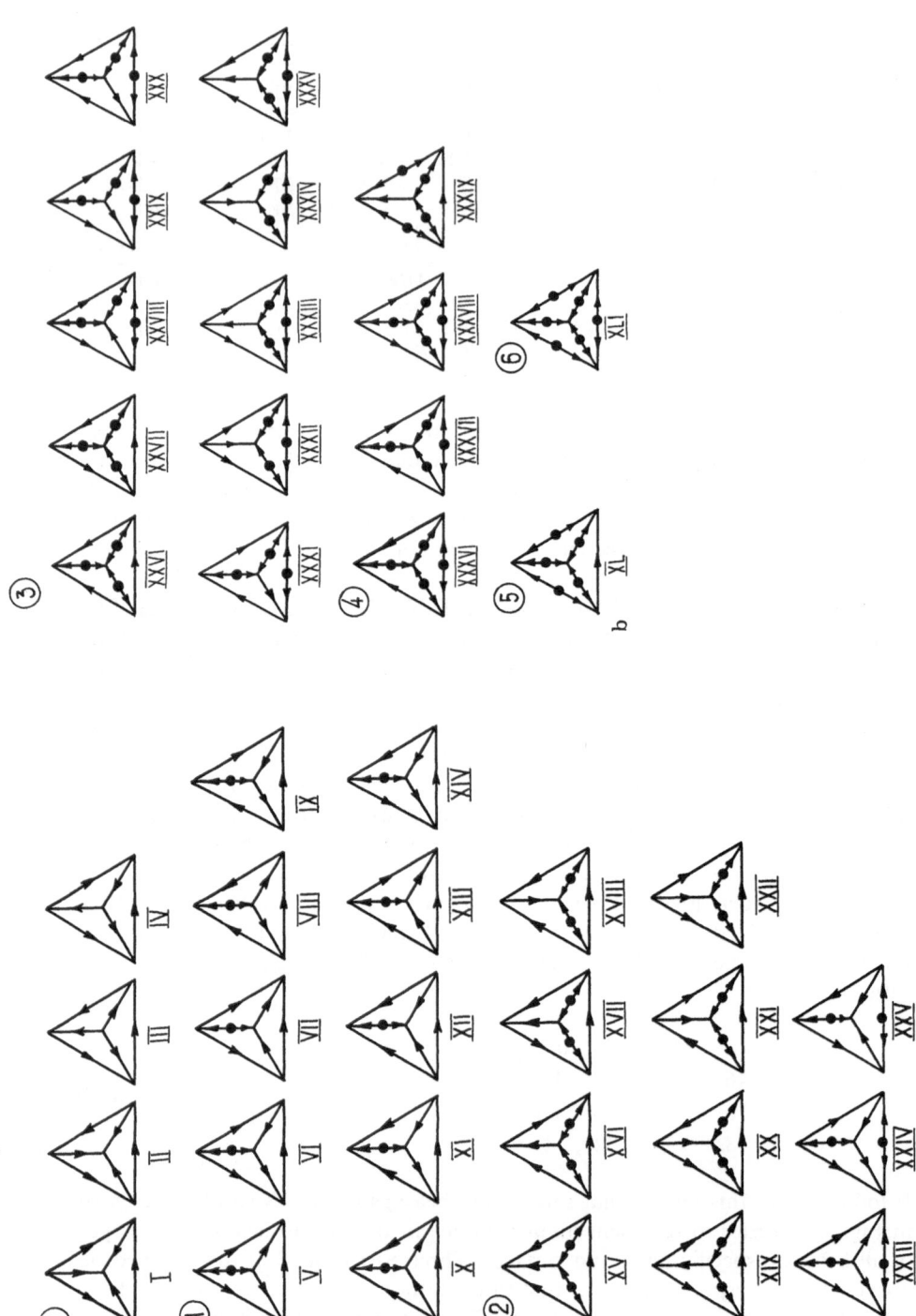

Fig. 9. Complete set of types of four-component systems. *Number in circle* denotes the number of binary azeotropes in such a system

ascertain the kinds (see Fig. 7) of the three-component subsystems (M_i, M_j, M_k) corresponding to the faces (ijk) of the tetrahedron:

Subsystem	(123)	(124)	(134)	(234)
Kind	13	10	10	7

It is sufficient for such an analysis to determine only the sign of the parameter λ of each binary azeotrope in all boundary 3-simplexes of the tetrahedron and to use Tables 5.2 and 5.3. Then by means of Table 5.6 we can establish that the binary azeotropes on the edges (12), (13), and (23) of the tetrahedron are the saddles of the 1-st order, and the binary azeotrope on the edge (14) is an unstable node. The sole triple azeotrope is the unstable node (see Fig. 5) and the saddle of 2-nd order (see Fig. 8) in the 3-simplex (123) and tetrahedron, respectively. The latter conclusion ensues from Table 5.6 since the value of the parameter $\lambda = D_4(1234)/D(123)$ calculated from the matrix (5.23) is negative. Having determined the types of all boundary SPs of the tetrahedron (and, consequently, the signs of their Poincare indices through Table 5.6), afterwards by means of formulae (5.22) one can calculate the value of the topological parameter δ. The result of such a calculation ($\delta = 0$) gives evidence to the absence of the azeotrope inside the tetrahedron. It is worth mentioning that for the above conclusion there is no need to calculate parameters ω_i for the four-component system under consideration. It gives evidence to the advantages of the topological approach based on the application of the "rule of azeotropy" (5.18).

On the basis of the information obtained within the framework of such an elementary analysis, one can build a phase portrait of the studied system. As it may be seen in Fig. 10, the tetrahedron has three stable SPs located at the apexes of its base. The basins of SPs are separated by the proper separatrix surfaces. The first of these passes through SPs of simplexes (14), (4), (23), (123), and the second one through SPs of simplexes (14), (13), (123), (12). Each basin consists of a single cell with the unstable SP which is common for all basins and located on the edge (14).

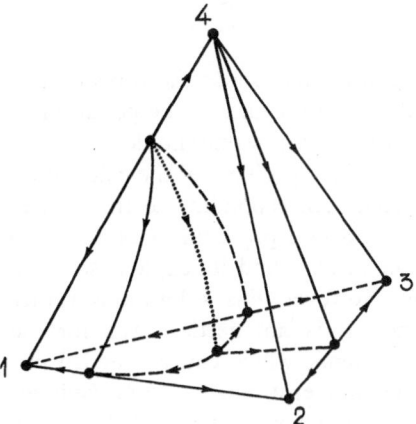

Fig. 10 Kind of phase portrait describing qualitatively the dynamics of tetrapolymerization of styrene, methyl methacrylate, 4-vinyl pyridine, and methyl acrylate, which is characterized by a set of kinetic parameters (5.23). In the diagram only specific lines (edges and separatrixes) are presented. Visible and invisible lines at a tetrahedron boundary are depicted as *solid* and *dashed lines*, respectively, and a line inside a tetrahedron is *dotted*

It is necessary to emphasize one principal peculiarity of the copolymerization dynamics which arises under the transition from the three-component to the four-component systems. While the attractors of the former systems are only SPs and limit cycles (see Fig. 5), for the latter ones we can also expect the realization of other more complex attractors [202]. Two-dimensional surfaces of torus on which the system accomplishes the complex oscillations (which are superpositions of the two simple oscillations with different periods) are regarded to be trivial examples of such attractors. Other similar attractors are fitted by the superpositions of few simple oscillations, the number of which is arbitrary. And, finally, the most complicated type of dynamic behavior of the system when $m \geq 4$ is fitted by chaotic oscillations [16], for which a so-called "strange attractor" is believed to be a mathematical image [206].

All the mentioned types of the nontrivial dynamic behavior are excluded for the systems where the reactivity ratios r_{ij} can be described by the expressions of the well-known Alfrey-Price "Q-e" scheme [20], and as a result they are to follow the simplified terminal model (see Sect. 4.6). In these systems, due to the relations $B_i(\vec{x})/B_j(\vec{x}) = a_{ji}/a_{ij}$ which holds for all i and j, the functions $\pi_i(\vec{x})$ according to relations (4.10) are the ratios of the homogeneous polynomials of degree 2. Besides, for the calculations of the coefficients α_k of Eq. (5.11) one can use the simple formulae presented in terms of determinants D_i and D [6, p. 265]. The theoretical analysis [202] leads to the conclusion that in such systems even the limited cycles are not possible and all azeotropes are certainly unstable. Hence any trajectory $\vec{x}(p)$ and $\vec{X}(p)$ when $p \to 1$ inevitably approaches the SP corresponding to the homopolymer the number of which can be from 1 to m. The set of systems obtained due to the classification within the framework of the simplified model essentially impoverishes in comparison with the general case of the terminal copolymerization model since some types of systems cannot be principally realized under the restrictions which the "Q-e" scheme puts on the reactivity ratios r_{ij}.

5.5 The Dynamics of Systems Described by Models Other than the Terminal one

The general formulae (5.1), (5.3), and (5.7) are still valid under the transition to the more complicated models described in Sect. 2. In the case of the penultimate model it concerns also the dynamic Eqs. (5.2) into which now one should substitute the dependence $\vec{X}(\vec{x})$ obtained after the solution of the problem of the calculation of the stationary vector $\vec{\pi}(\vec{x})$ of the Markov chain corresponding to this model. Substituting the function $X_1(x_1)$ obtained via the above procedure (see Sect. 3.1) into Eq. (5.2) for the binary copolymerization we can find its explicit solution expressed through the elementary functions. However, this solution is rather cumbersome and has no practical importance. It is not needed even for the classification of the dynamic behavior of the systems, which can be carried out only via analysis of Eq. (5.5) by determining the number $n = 0, 1, 2, 3$ of the inner azeotropes in the 2-simplex [14]. The complete set of phase portraits of the binary copolymerization described by the penultimate model in addition to the kinds snown in (5.8) involves two more [14, 18]:

$$1 \bullet \!\!\longrightarrow\!\! \circ \!\!\longleftarrow\!\! \bullet \!\!\rightarrow\!\! \circ \; 2 \; (III) \; 1 \; \circ \!\!\leftarrow\!\! \bullet \!\!\rightarrow\!\! \circ \!\!\leftarrow\!\! \bullet \!\!\rightarrow\!\! \circ \; 2 \; (IV) \qquad (5.24)$$

which cannot be realized within the framework of the terminal model. It follows from pictures (5.8) and (5.24) that each kind of phase portrait is characterized by its number n of inner azeotropes. Since the value of n for any concrete system can be found by means of the elementary analysis of expressions (5.5) and (5.6), the problem of determining the system kind becomes rather simple.

Within the framework of the model taking into account the reactions (2.5) of the complex formation between two monomers, one cannot use Eq. (5.2) if the fractions of the free and complexed forms of the monomer are comparable. In the case of systems for which the participation of the monomer units in complexing is less pronounced in comparison with monomers, the conditions of stoichiometry and equilibrium:

$$M_1 + \bar{M}_1 + M_{12} = M_1^0, \qquad M_2 + \bar{M}_2 + M_{12} = M_2^0,$$
$$M_{12} = kM_1M_2 \qquad\qquad (5.25)$$

allow one to obtain the following kinetic equations:

$$-\frac{dM_1}{dp} = \frac{(1 + kM_1)X_1 - kM_1X_2}{1 + k(M_1 + M_2)}, \qquad M_1(0) = M_1^0,$$

$$\qquad\qquad\qquad\qquad\qquad\qquad\qquad\qquad\qquad\qquad (5.26)$$

$$-\frac{dM_2}{dp} = \frac{(1 + kM_2)X_2 - kM_2X_1}{1 + k(M_1 + M_2)}, \qquad M_2(0) = M_2^0$$

which describe the drift of monomer concentration with conversion. This set of equations is of closed form since the dependence of the instantaneous copolymer composition X_1, X_2 on M_1, M_2, M_{12} is known from the statistical consideration presented in Sect. 3.2. If $k(M_1 + M_2) \ll 1$, i.e., when the current complex concentration is much lower than the monomer concentration, Eqs. (5.26) are reduced to the form (5.2). However, the copolymer composition $\vec{X}$ is now determined not only by the monomer feed composition $\vec{x}$ but also by the conversion p since $\vec{X}$ depends not only on the ratio of monomer concentrations but on their absolute values of M_1 and M_2. Therefore due to the explicit dependence of the right-hand part of Eq. (5.2) on p one cannot obtain its analytical solution. The above case is of some practical interest since despite the low complex content as compared to the free monomers it can have a better ability to be added to the radical than monomers do. As a result, the fractions of monomers which have been introduced into the polymer chain as single molecules and as monomer pairs are comparable.

5.6 Composition Inhomogeneity of the Multicomponent Copolymerization Products

As already noted in Sect. 5.1 for the copolymers produced at high conversion, one can neglect the "instantaneous" component of the composition distribution and then calculate it by means of formula (5.3). As follows from this formula, the form of this distribution at a given initial monomer feed composition $\vec{x}^0$ is determined only by the character of the trajectory $\vec{X}(p)$ in the m-simplex of the instantaneous copolymer composition, passing through the point $\vec{X}^0 \equiv \vec{X}(\vec{x}^0)$. The variation range of the components X_i of vector $\vec{X}$ on this trajectory determines the intervals of the copolymer composition values, within which its function of the composition distribution is different from zero. When these intervals are found to be rather narrow (e.g., when the initial composition $\vec{x}^0$ is located near the stable SP within its basin) one can make a certain conclusion concerning the formation in this case of the compositionally uniform copolymers. If these intervals are rather wide, one cannot derive such a conclusion without any additional theoretical analysis since the composition distribution for the various systems may be both wide and narrow depending on the number, form, and mutual arrangement of its possible modes. Such an analysis to a great extent is based on the dynamic consideration of copolymerization.

The binary systems have two possible qualitatively different forms of the composition distribution of the products of the complete conversion ($p = 1$) shown in Fig. 4. The bimodal distribution is realized when the initial system composition $\vec{x}^0$ is inside the basin of the stable SP corresponding to the homopolymer of the monomer M_s and the reactivity ratio of the corresponding radical $r_s < 0.5$. In the opposite case when $0.5 < r_s < 1$ the composition distribution is unimodal. These conclusions derived at first [169–170] from the expressions (5.4) can also be obtained as a particular case from the general results [6, 134] of the analysis of the trajectory behavior near SP for the copolymerization of any number m of monomers.

Since the stable SPs are usually located at the apexes of the m-simplex $x_i^* = \delta_{is}$, it is important to know which products are formed in such SP vicinity. It is determined by the number ν of the reactivity ratios r_{si} exceeding 0.5 [202]. When $\nu = 0$, at $p \to 1$ in the composition distribution arises a mode, the maximum of which corresponds to the homopolymer M_s. When $1 \leq \nu \leq (m - 2)$, it was found that the product of the copolymerization at its final stage is random ($\nu + 1$)-component copolymer of monomer M_s with those ν monomers M_i whose parameters $r_{si} > 0.5$. At last, in the case $\nu = m - 1$ when all $r_{si} > 0.5$, a new mode of composition distribution does not arise at $p \to 1$ and at high conversion the formation of the random copolymer containing all m types of monomer units occurs. In the latter case the stable SP under consideration, referred to as regular, does not really contribute to the composition inhomogeneity of the copolymer.

When the stable SP is azeotrope, necessary and sufficient conditions for its regularity are the following: the real parts Re λ_i of all roots λ_i of the characteristic equation (5.11) are to exceed (-1) [6, 134]. When inverse non-equalities Re $\lambda_i < -1$ hold for all $i = 1, ..., m - 1$, at the conversions close to the complete one in the composition distribution, we can observe a mode, the maximum of which corresponds to the azeotropic composition of the considered SP.

In studying the composition inhomogeneity of the multicomponent copolymerization products we need to devote special attention to those cases where the monomer feed composition $\vec{x}^0$ is located near any specific trajectory or surface, which usually separate the basins of the various stable SPs. Since such a trajectory approaches the unstable SP $\vec{x}_u^*$, the system may delay in the vicinity of the trajectories close to the specific one up to the almost complete conversion and only then goes to the proper stable SP $\vec{x}_s^*$. Consequently, the composition inhomogeneity of the copolymer produced on such trajectories is determined by the values of the roots λ_i of Eq. (5.11) not for the point $\vec{x}_s^*$ but for $\vec{x}_u^*$ in the vicinity of which at $p \to 1$ the corresponding distribution mode may be formed. It also concerns the trajectories starting near the specific surface, since the trajectories located on it may enter the unstable SP, through which this separatrix surface passes. For example, in the case of the three-component systems of kind 11 presented in Fig. 7 at initial compositions $\vec{x}^0$ located near the separatrix, which enters a saddle point at the bottom edge (12), the degree of composition inhomogeneity of terpolymers will depend upon the value of the root λ_2 (5.15) of the characteristic equation (5.11) of this saddle. When the regularity conditions $\mathrm{Re}\,\lambda_2 > -1$ hold, one can expect the formation of a uniform copolymer.

The results of the above analysis are not restricted within the framework of the simplest kinetic scheme (2.8) and allow one to consider similarly the copolymerization described by the more complex models. In particular, in the case of the penultimate model (2.3), systems which have the stable inner azeotrope are possible. Hence when the initial compositions $\vec{x}^0$ are located inside its basin, one should know for certain whether this azeotrope is regular or not. If the regularity condition $\mathrm{Re}\,\lambda_1 > -1$ in the following form [14, 18]:

$$(2 - r_1)\, r_1'(x_1^*)^3 + (1 + 2r_1'r_2')\, x_1^*x_2^* + (2 - r_2)\, r_2'(x_2^*)^3 > 0 \qquad (5.27)$$

holds, the composition distribution is unimodal. But in the case where the non-equality inverse to (5.27) one is hold, at high conversions the second mode arises which has a maximum corresponding to the composition of the stable inner azeotrope. For the systems (5.24) this kind of composition distribution is allowed which is not realized in the case of the terminal model. For instance, this distribution can have two maxima, no one of which corresponds to homopolymer.

While the products of the binary copolymerization can be characterized only by the two kinds of the composition distribution shown in Fig. 4, in the case of $m \geq 3$ monomers, the number of the different kinds of the one-component distribution (5.3) noticeably increases. The mode of the infinite amplitude in the composition distribution of $\bar{M}_i$ units of monomer M_i arises at the conversions p^e where the function $X_i(p)$ reaches its extreme values. The locus of these points, where $dX_i/dp = 0$ being the manifold of the $(m - 2)$-dimension, is located inside the m-simplex and described by the following relationships:

$$X_i = S_i / \sum_{j=1}^{m} S_j, \qquad S_i \equiv \sum_{j=1}^{m} X_j \frac{\partial \Delta_i}{\partial x_j}, \qquad (5.28)$$

$$S_i = \pi_i B_i \sigma_i + x_i \sum_{j=1}^{m} (B_i a_{ij} + \sigma_i B_{ij}) \pi_j, \qquad (5.29)$$

the latter of which is valid only within the framework of the terminal model. The manifold (5.28) will certainly pass through SP since at $X_i = x_i$ we have $S_i = m\Delta_i$.

In the case of a terpolymerization, the relationship (5.28) for each monomer M_i gives a line, at the intersection of which by the trajectory the amplitude of the proper mode of the i-th composition distribution (5.3) turns into infinity. The character of the mutual arrangement of this singular i-th line and separatrixes inside the Gibbs-Roozeboom triangle allows one to draw some conclusions concerning the dependence of the number of modes of the i-th composition distribution (5.3) on the copolymer composition and reactivity ratios. One may also note that the considered i-th singular line is the locus of the points where the tangents to the trajectories $X(p)$ inside the triangle are directed parallelly to the edge, located in front of the i-th apex. For the four-component copolymerization, the relationship (5.28) determines for each monomer M_i a proper singular surface which in the tetrahedron has the same meaning as the i-th singular line has in the triangle for terpolymer.

The composition distributions (5.3), besides the ones considered above, can involve the modes of the finite amplitude, for which the locations of maxima as well as the boundary points separating such modes are some $(m - 2)$-dimensional manifolds in the m-simplex.

At the end of the present section we can emphasize that the results of the above abstract mathematical analysis are very effective for the prediction of some properties of multicomponent copolymers forming in the course their synthesis. This will be demonstrated in Sect. 7 conformably to particular copolymers based on commercial monomers.

6 Discrimination of the Kinetic Models and Estimation of their Parameters

6.1 General Strategy

When the copolymerization is carried out under real conditions, each researcher is to answer a question which kinetic model is preferable for the proper description of the experimental data. One should also know the validity of the model under consideration, the numerical values of its parameters, and the expected accuracy of the calculated copolymer characteristics predicted within the framework of this model. Modern experimental methods for analyzing the copolymer composition and structure allows one to solve all the above problems quite reliably for any particular system [207].

Even in the early report [208] it was stated that in order to draw a well-founded conclusion concerning the applicability of this or that kinetic model of copolymerization one needs not only the data on the copolymer composition but also additional information on the macromolecular chemical structure. In practice this procedure can be presented as follows. At first one should determine at low conversions the dependence of the copolymer composition X_1 on the monomer feed mixture composition x_1 over a wide range of its variation. It can be easily done using such techniques as elementary analysis, nuclear magnetic resonance spectroscopy (NMR), radioactive labels, etc. Then according to the dependence of X_1 on x_1 one can estimate those relative reactivity ratios (2.2) which fit the experimental data best. Knowing r_1 and r_2 estimates, one can calculate fractions of different sequences of the monomer units and compare them to the experimental data obtained by means of spectroscopy, pirolytic chromatography, or other techniques. An example of such a complex approach to the copolymerization in the systems where one monomer is vinyl chloride are presented in Figs. 11–14. The reactivity ratios in this case were estimated from the data on chlorine content in the copolymer, and the theoretical curves were calculated within the terminal model. Figures 11–14 bear convincing evidence that the above model can be successfully applied to describe the microstructure of the copolymers under study [209].

In those cases where for the homophase processes such a coincidence does not occur one should apply a similar complex approach to describe the experimental data using kinetic models other than the terminal one. When such an approach is realized in practice the trustworthiness of the final conclusions is determined

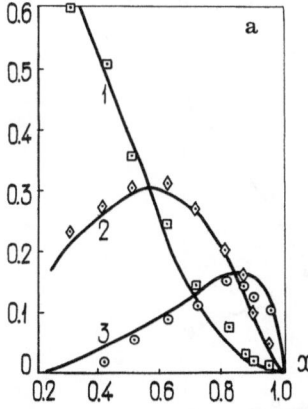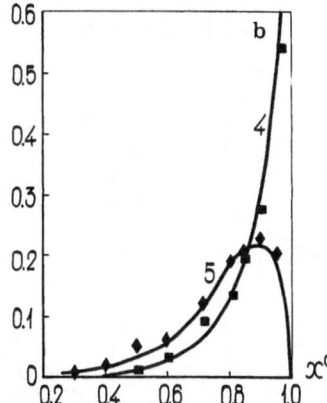

Fig. 11a, b. Theoretical *(curves)* and experimental *(points)* dependences of triads $P(\bar{M}_2\bar{M}_2\bar{M}_2)$ *(1)*, $P(\bar{M}_1\bar{M}_2\bar{M}_2)$ *(2)*, $P(\bar{M}_1\bar{M}_2\bar{M}_1)$ *(3)* **(a)**, and tetrads $P(\bar{M}_1\bar{M}_1\bar{M}_1\bar{M}_1)$ *(4)*, $P(\bar{M}_1\bar{M}_1\bar{M}_1\bar{M}_2)$ *(5)* **(b)** on monomer feed composition for the copolymer of vinylidene chloride M_1 with methyl methacrylate M_2 prepared at low conversions [209]

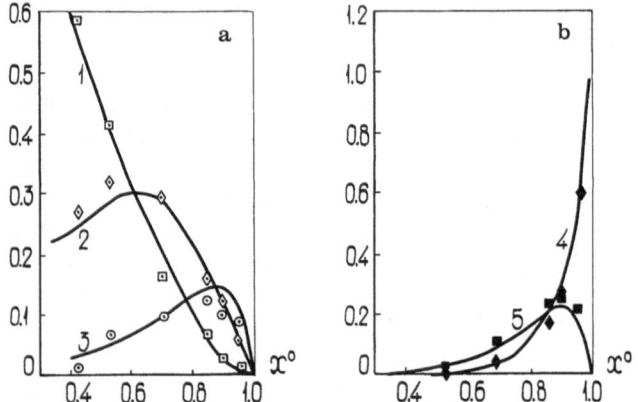

Fig. 12a, b. Plots similar to those shown in Fig. 11 but for the copolymer of vinylidene chloride M_1 and benzyl methacrylate M_2 [209]

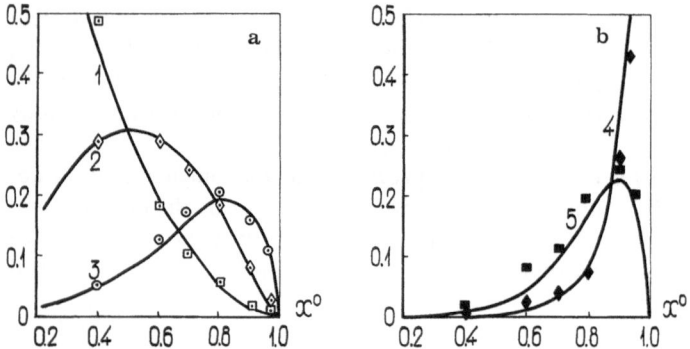

Fig. 13a, b. Plots similar to those shown in Fig. 11 but for the copolymer of vinylidene chloride M_1 and chloro methacrylate M_2 [209]

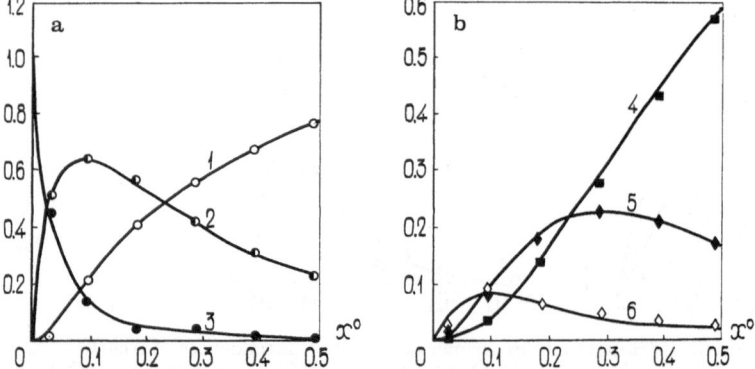

Fig. 14a, b. Plots similar to those shown in Fig. 11 plus the fractions of tetrads $P(\bar{M}_2\bar{M}_1\bar{M}_1\bar{M}_2)$ (6) for the copolymer of vinylidene chloride M_1 with vinyl acetate M_2 [209]

by a number of factors. These factors involve, for example, the estimation accuracy of the kinetic parameters which in its turn is dependent either on the measurement errors due to the determination of the copolymer composition or on the errors arising from the treatment of the experimental data. The knowledge of the accuracy level of the reactivity ratios estimates is of great importance since it allows to define the confidence limits of the probabilities of the different sequences. When these experimental values do not fall within such limits, one may conclude for certain that this particular model fails to describe this system properly. Such a complex analysis permits one in a scientifically grounded way to approach the problem how to select a kinetic model adequate to the system under consideration.

6.2 Reactivity Ratios Estimation Based on Copolymer Composition Data

The problem of the computation accuracy of these kinetic parameters is dependent first of all on the validity of the copolymer composition determination. As a criterion here one may use the closeness to each other of the values of this composition obtained via the different experimental methods. It is possible to judge about the degree of such a closeness using Tables 6.1 and 6.2 where the data on both chemical analysis and spectroscopy are presented. One can see that, as for the considered cases, the different experimental methods provide quite close values of the copolymer compositions within the accuracy in the range of 5%. Authentic evidence concerning the feasibility to reach such a degree of accuracy is furnished by the data on copolymer composition obtained via independent methods in the different systems, for instance, under the copolymerization of p-chlorstyrene with methyl acrylate [32], of 4-methylstyrene with methyl methacrylate or acrylonitrile [213], and also of styrene with acrylic or methacrylic acids [214].

Another reason for errors of the reactivity ratio values are an exactitude in the course of the treatment of the experimental data using the differential or integrated form of the copolymer composition equation. In the first case, the dependence of $X(x^0)$ on the monomer feed composition x^0 experimentally determined at low conversions is used. In the second case, one should use the data on the dependence of the copolymer composition on conversion p or the current values of x under the measurements of p.

A number of well-known procedures of r_1, r_2 estimations within the framework of the terminal copolymerization model share the following general expression:

$$y = \frac{z^0(1 + r_1 z^0)}{r_2 + z^0}, \quad \text{where} \quad z^0 = \frac{x_1^0}{x_2^0}, \quad y = \frac{X_1}{X_2} \tag{6.1}$$

which allows one, knowing the experimental dependence y on z^0, to calculate the reactivity ratios. The difference between these procedures is due only to the way of writing down expression (6.1).

Table 6.1 Copolymer composition X_2 versus monomer feed composition x_2^0 in homogeneous copolymerization of styrene M_1 and methyl methacrylate M_2 T = 60 °C at low conversions (less than 10%) determined by means of NMR and elementary carbon analysis. (Abbreviations NMR1 and NMR2 correspond to spectra with resonating proton located in styrene and methyl methacrylate unit, respectively)

In bulk [210]				In benzene [211]			In chlorobenzene [211]		
$x_2^0 \cdot 100$	$X_2 \cdot 100$			$x_2^0 \cdot 100$	$X_2 \cdot 100$		$x_2^0 \cdot 100$	$X_2 \cdot 100$	
	NMR1	NMR2	CA		NMR2	CA		NMR2	CA
14.2	24.0	24.0	—	20.0	26.8	25.9	20	29.1	29.2
29.9	38.3	37.3	36.7	30.0	34.8	36.2	30	38.4	34.8
39.2	44.5	44.9	44.0	40.0	42.7	43.0	40	43.0	43.1
50.8	52.2	51.2	52.9	50.0	48.6	48.6	50	51.4	50.6
60.6	59.1	60.3	59.3	60.0	55.1	55.4	60	56.9	57.3
69.8	64.1	65.0	65.9	70.0	62.2	62.7	70	64.0	64.1
70.9	65.3	65.5	—	80.0	71.6	69.3	80	72.6	73.6
78.6	71.9	74.6	—	—	—	—	—	—	—
88.3	81.1	82.7	—	—	—	—	—	—	—
r_1	0.45 ± 0.10			0.52 ± 0.03			0.48 ± 0.05		
r_2	0.50 ± 0.05			0.47 ± 0.04			0.50 ± 0.03		

Table 6.2 Copolymer composition X_1 versus monomer feed composition x_1^0 of bulk radical copolymerization of glycidyl methacrylate M_1 and styrene M_2 ($T = 60\,°C$, $p < 0.1$). In three columns the experimental data obtained [217] by means of chemical analysis of epoxy group content (EA), Infra Red (IR) and Proton Nuclear Magnetic Resonance (NMR) spectra are presented

x_1^0	Copolymer composition X_1		
	EA	IR	NMR
0.2	0.34	0.36	0.36
0.4	0.52	0.51	0.50
0.6	0.65	0.62	0.63
0.8	0.78	0.80	0.79

The "intersection procedure" put forth in the pioneer report by Mayo-Lewis (ML) [19] assumes expression (6.1) to adopt the following form:

$$r_2 = z^0[(1 + r_1 z^0)/y - 1]. \tag{6.2}$$

Each experiment which is characterized by a pair of (z^0, y)-values gives a straight line in the (r_1, r_2)-plane, and the region of the intersection of the lines corresponding to the experiments performed for the different monomer feed compositions provides unknown values of the parameters r_1 and r_2.

In the other well-known procedure proposed by Finemann-Ross (FR) [215], the expression (6.1) is used in one of the following two forms:

$$1)\ G = r_1 F - r_2, \quad 2)\ (G/F) = r_1 - r_2(1/F) \tag{6.3}$$

where the parameters G and F are defined as:

$$G = z^0(y - 1)/y, \quad F = (z^0)^2/y \tag{6.4}$$

and have certain values for each experimental point (z^0, y). Consequently, the slopes of the plots G vs F or (G/F) vs $(1/F)$ and their intersection points with the axes give the needed values of r_1 and r_2. However, this method suffers from a number of disadvantages since the resultant values are dependent upon the mode of monomer indexing. Below, two other procedures will be discussed which are free from the above drawbacks.

The first procedure by Yezrielev, Brokhina, and Roskin (YBR) [216] is a result of the symmetrical transformation of expression (6.1) to give:

$$r_2 F^{-1/2} = r_1 F^{1/2} - y^{1/2} + y^{-1/2}. \tag{6.5}$$

This method is not subject to reindexing errors and gives a valid estimate of the confidence limits while determining the parameters r_1 and r_2.

The second of the above-mentioned procedures proposed by Kelen and Tudos (KT) [217] is based on the transformation of expression (6.1) into:

$$\eta = \xi(r_1 + r_2/\alpha) - r_2/\alpha \qquad (6.6)$$

where the transformed variables η and ξ are defined as:

$$\eta = G/(\alpha + F), \qquad \xi = F/(\alpha + F) \qquad (6.7)$$

and the symmetry parameter $\alpha = (F_{min}/F_{max})^{0.5}$ is expressed through the minimum and maximum values of F. The plot of η vs ξ is a straight line which on extrapolation to $\xi = 0$ and $\xi = 1$ gives according to expression (6.6) the values of the reactivity ratios.

One more method which is not subject to reindexing errors and which allows, like the two above procedures, to determine the confidence limits of the reactivity ratios was proposed by Joshi and Joshi (JJ) [218]. This method being a further modification of the "intersection procedure" (IP) is free of its intrinsic disadvantages.

Other graphical [219, 220] and computing [221–223] procedures of r_1, r_2 estimation also deserve to be mentioned despite their rather limited application. Using the computer one can estimate the reactivity ratios via minimization of the modulus of the deviation of the observed copolymer composition from the computed one. Nevertheless, this procedure performed in Refs. [219–223] has the same disadvantages as any other aforementioned linear least-squares procedures.

A quarter of a century ago Behnken [224] as well as Tidwell and Mortimer [225] pointed out that the linearization transforms the error structure in the observed copolymer composition with the result that such errors after transformation have no longer zero mean and constant variances. It means that such transformed variables do not meet the requirements for the least-squares procedure. The only statistically accurate means of estimation of the reactivity ratios from the experimental data is based on the non-linear least-squares procedure. An effective computing program for this purpose has been published by Tidwell and Mortimer (TM) [225]. Their method is considered to be such a modification of the curve-fitting procedure where the sum of the squares of the difference between the observed and computed polymer compositions is minimized.

Despite certain intrinsic statistical disadvantages the linear least-square procedures can sometimes provide satisfactory r_1, r_2 estimates when the copolymerization experiment is properly designed. Instead of conducting all the measurements over the random range of the initial monomer feed composition, one should carry them out only at the two following values:

$$x_1' = 2/(2 + r_1^*), \qquad x_1'' = r_2^*/(2 + r_2^*) \qquad (6.8)$$

where r_1^* and r_2^* are the approximated values of the preliminary estimated reactivity ratios [225]. Such a strategy in comparison with the traditional one is believed to provide more accurate r_1, r_2 determination.

The thorough treatment of the experimental data does allow one to obtain reliable values of the reactivity ratios. The results of such a treatment are presented in Table 6.3 for some concrete system let us form a notion about an accuracy of the reactivity ratios estimations. The detailed analysis of such a significant problem in the case of the well-studied copolymerization of styrene with methyl methacrylate is reported in Ref. [227]. Important results on the comparison of the precision of r_1, r_2 estimates by means of different methods are presented by O'Driscoll et al. [228]. Such a comparison of six well-known linear least-squares procedures [215–218, 222, 223] with the statistically correct non-linear least-squares method leads to the conclusion that some of them [216, 217, 222] can provide rather precise r_1, r_2 estimates when the experiment is properly planned.

Numerous reports are available [19, 229–248] on the development and analysis of the different procedures of estimating the reactivity ratio from the experimental data obtained over a wide range of conversions. These procedures employ different modifications of the integrated form of the copolymerization equation. For example, "intersection" [19, 229, 231, 235], (KT) [236, 240], (YBR) [235], and other [242] linear least-squares procedures have been developed for the treatment of initial polymer composition data. Naturally, the application of the non-linear procedures allows one to obtain more accurate estimates of the reactivity ratios. However, majority of the calculation procedures suffers from the fact that the measurement errors of the independent variable (the monomer feed composition) are not considered. This simplification can lead in certain cases to significant errors in the estimated kinetic parameters [239]. Special methods [238, 239, 241, 247] were developed to avoid these difficulties. One of them called "error-in-variables method" (EVM) [239, 241, 247] seems to be the best. EVM implies a statistical approach to the general problem of estimating parameters in mathematical models when the errors in all measured variables are taken into account. Though this method requires more information than do ordinary non-linear least-squares procedures, it provides more reliable estimates of r_1 and r_2 as well as their confidence limits.

Certain achievements in the accurate estimation of the reactivity ratios, particularly where the integrated form of the copolymerization equation is

Table 6.3 Summary of reactivity ratios determined by various methods for the bulk copolymerization of styrene with methyl α-hydroxymethyl acrylate at T = 80 °C [226]

Methods	r_1	r_2
FR	0.37	0.34
JJ	0.346 ± 0.162	0.363 ± 0.069
YBR	0.349 ± 0.016	0.326 ± 0.032
KT	0.326 ± 0.046	0.325 ± 0.047
TM	0.330	0.326

employed, are connected to a great extent with the introduction of computers. A number of special computer programs have been developed to estimate r_1 and r_2 [230, 231, 235] and to determine "Q" and "e" values of the Alfrey-Price ("Q-e") scheme [249–251]. Discussions concerning some items of design of experiment and as well as special computer problems arising during the estimation of the reactivity ratios are reported in Refs. [252–255].

When r_1, r_2 values are rather close to unity, one can use for their estimation the so-called "approximation method" [225, 256–258]. Its idea is based on the fact that if the copolymerization is carried out at low concentrations of one of the monomers, the instantaneous composition of the copolymer depends only on one reactivity ratio. In this case the composition equation in both differential and integrated forms is fairly simple.

It is quite obvious that in the framework of the present review the author has no possibility to consider circumstantially all the problems concerning the estimation of r_1 and r_2 from the experimental measurements of copolymer composition. All these problems have been already discussed in numerous reviews and monographs [259–261, 239, 207 p. 105, 262 p. 275, 263–265].

As mentioned above, the equations of the multicomponent copolymer composition within the framework of the general terminal model contain only the reactivity ratios r_{ij} experimentally estimated under the copolymerization of different pairs of monomers M_i and M_j. For example, studying three binary systems one can determine six kinetic parameters r_{ij} that completely characterize the terpolymer composition. However, some authors [266–269, 261, 131, 270] propose to determine these values immediately from the experimental data obtained under the ternary copolymerization at low conversions. It is worth mentioning that to estimate in this way the reactivity ratios, the different calculation methods were applied just as it has been already done for the binary systems. Both, the simple linear procedures [266, 267, 131] as well as the far more complicated non-linear ones [268–270] were used. The discussion on the early works concerning these problems and the corresponding computer programs is reported in a review [261]. The generalization of the well-known non-linear least-squares procedure (EVM) for the ternary systems was put forth by the authors of Ref. [270]. A similar generalization of the trivial approximation method is presented in Ref. [256].

The kinetic copolymerization models, which are more complex than the terminal one, involve as a rule no less than four kinetic parameters. So one has no hope to estimate their values reliably enough from a single experimental plot of the copolymer composition vs monomer feed composition. However, when in certain systems some of the elementary propagation reactions are forbidden due to the specificity of the corresponding monomers and radicals, the less number of the kinetic parameters is required. For example, when the copolymerization of two monomers, one of which cannot homopolymerize, is known to follow the penultimate model, the copolymer composition is found to be dependent only on two such parameters. It was proposed [26, 271] to use this feature to estimate the reactivity ratios in analogous systems by means of the procedures similar to ones outlined in this section.

6.3 Application of NMR and Discrimination of the Kinetic Models of Copolymerization

The application of NMR spectroscopy data to estimate the reactivity ratios is regarded to be very promising [272]. The Q and e values of the Alfrey-Price scheme may be immediately calculated analyzing the shifts of the corresponding bands in carbon-NMR spectra. Such data obtained for more than fifty pairs of monomers are tabulated in Ref. [273]. A quite different method based on the application of the trivial expressions:

$$r_1 = \frac{2x_2^0 P(\bar{M}_1\bar{M}_1)}{x_1^0 P(\bar{M}_1\bar{M}_2)}, \qquad r_2 = \frac{2x_1^0 P(\bar{M}_2\bar{M}_2)}{x_2^0 P(\bar{M}_1\bar{M}_2)} \tag{6.9}$$

is proposed in Ref. [274]. It permits one to express the reactivity ratios through both the initial monomer feed composition x_1^0 and the fractions of the dyads $P(U_2) \equiv P(\bar{M}_i\bar{M}_j)$. The latters are determined by analyzing NMR spectra of polymers obtained at low conversions. The invariability of the values of r_1 calculated according to formulae (6.9) over a wide composition x_1^0 range may serve as an objective criterion of the applicability of the terminal copolymerization model to the system under consideration.

When the above-mentioned independence of r_i on composition is not the case, it is quite necessary to study different possibilities of the description of the copolymerization in a given system by means of more complicated models. For instance, to establish the applicability of the penultimate model for the copolymers produced at low conversions, one may use the following relations [275]:

$$r_1 = \frac{2x_2^0 P(\bar{M}_1\bar{M}_1\bar{M}_1)}{x_1^0 P(\bar{M}_1\bar{M}_1\bar{M}_2)}, \qquad r_1' = \frac{x_2^0 P(\bar{M}_1\bar{M}_1\bar{M}_2)}{2x_1^0 P(\bar{M}_2\bar{M}_1\bar{M}_2)}, \tag{6.10}$$

$$r_2 = \frac{2x_1^0 P(\bar{M}_2\bar{M}_2\bar{M}_2)}{x_2^0 P(\bar{M}_1\bar{M}_2\bar{M}_2)}, \qquad r_2' = \frac{x_1^0 P(\bar{M}_1\bar{M}_2\bar{M}_2)}{2x_2^0 P(\bar{M}_1\bar{M}_2\bar{M}_1)}, \tag{6.11}$$

which are analogous to expressions (6.9). By means of the above expressions, the parameters (2.4) of the penultimate model are immediately expressed through x_i^0 and fractions of the different triads $P(U_3) \equiv P(\bar{M}_i\bar{M}_j\bar{M}_k)$. If the substitution of the values of these fractions obtained by means of spectroscopy into the relations (6.10), (6.11) yields the reactivity ratios (2.4), independent on the initial monomer feed composition one may conclude that the copolymerization penultimate model is valid and simultaneously estimate its parameters.

Such a procedure was successfully employed for the copolymerization of acrylonitrile with methyl methacrylate [275] or methacrylic acid [276]. In the former case, the copolymerization was carried out at 40 °C in bulk (I) and in tetrahydrofurane (II) and the following values of the kinetic parameters were obtained [275]:

$$\text{(I)} \quad r_2 = 1.26 \pm 0.01 , \qquad r_2' = 1.28 \pm 0.03 ,$$
$$\text{(II)} \quad r_2 = 1.28 \pm 0.01 , \qquad r_2' = 1.35 \pm 0.05 . \tag{6.12}$$

In the latter case [276] all four parameters (2.4) for two different solvents were estimated (see Table 6.4). On the basis of the above data, the reliability of which is high enough, one may draw some important conclusions. In the former case no influence of the environment on the reactivity ratio was observed; in the latter

Table 6.4 Values of reactivity ratios calculated according to Eqs. (6.10) and (6.11) used in Ref. [276] to treat the data on products of copolymerization of acrylonitrile with methacrylic acid in solution of dimethyl sulfoxide (I) and its equimolar aqueous mixture (II). The conversion in all cases does not exceed 7%

x_2^0	I				II			
	r_1	r_1'	r_2	r_2'	r_1	r_1'	r_2	r_2'
0.15	0.34	0.31	—	—	0.21	0.19	—	—
0.20	0.36	0.32	—	0.84	0.22	0.19	—	1.38
0.25	0.36	0.34	—	—	0.24	0.22	—	—
0.30	0.38	0.35	0.65	0.89	0.24	0.21	0.96	1.32
0.40	0.37	0.32	—	—	0.23	0.19	—	—
0.50	0.34	0.35	0.67	0.83	0.23	0.20	0.99	1.36
0.60	0.35	0.35	—	—	—	0.23	—	—
0.70	—	0.33	0.65	0.84	—	—	0.98	1.35
0.80	—	—	0.64	0.82	—	—	—	—

case this influence was found to be rather noticeable. Besides, as it can be seen from the data presented in (6.12) the reactivity ratio of methyl methacrylate radical is practically independent on the type of the penultimate unit, $(r_2' = r_2)$, as for radical of methacrylic acid such a dependence is pronounced fairly well, $(r_2' \neq r_2)$. Considering the acrylonitrile radical, the values of r_1' and r_1, as it follows from Table 6.4, in each solvent are the same within the accuracy of the experiment, $(r_1' = r_1)$. It means that in both solvents the influence of the penultimate unit on the reactivity ratio of this radical is not observed when it adds acrylonitrile and methacrylic acid.

One may also mention one simple and convenient procedure [277–279], which allows knowing the fractions of the triads, $P_{ik}^{(j)} \equiv P(\bar{M}_i \bar{M}_j \bar{M}_k)/P(\bar{M}_j)$, centered by the monomer unit $\bar{M}_j$ to establish the adequacy of the terminal model and simultaneously to estimate its parameter r_j. This procedure involves the following

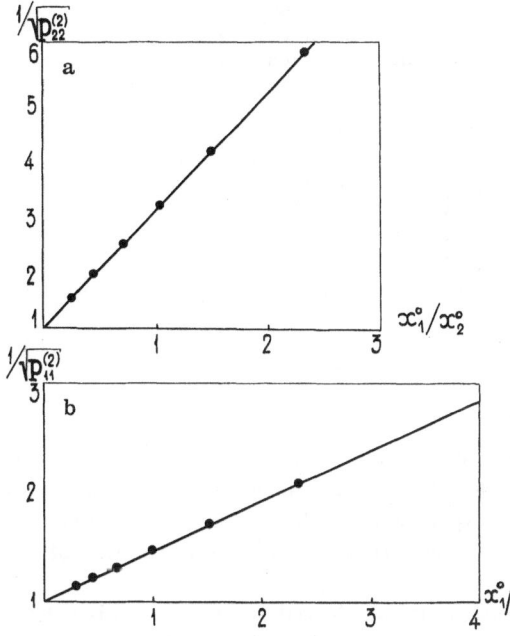

Fig. 15 Verification (according to Eqs. (6.1)) of the applicability of the terminal model on the basis of NMR data of copolymer of styrene with methyl methacrylate [279]

relations:

$$\frac{1}{\sqrt{P_{11}^{(1)}}} = 1 + \frac{x_2^0}{r_1 x_1^0}, \qquad \frac{1}{\sqrt{P_{22}^{(1)}}} = 1 + \frac{r_1 x_1^0}{x_2^0},$$

$$\frac{1}{\sqrt{P_{22}^{(2)}}} = 1 + \frac{x_1^0}{r_2 x_2^0}, \qquad \frac{1}{\sqrt{P_{11}^{(2)}}} = 1 + \frac{r_2 x_2^0}{x_1^0} \qquad (6.13)$$

which are to hold for the systems described within the framework of the above model. Therefore, the condition of the model applicability is the linear dependence of $1/\sqrt{P_{ii}^{(j)}}$ on the ratio of the initial mole monomer fractions in the reaction system, corresponding to relations (6.13). As it can be seen in Fig. 15 the experimental data, obtained by means of NMR spectroscopy [280] (see Table 6.5), fit the straight lines in the coordinates of the third and forth relations (6.13). The slopes of these straight lines yield the values of r_2 which are equal to 0.49 and 0.46, respectively. So, an excellent correlation with the values of the reactivity ratios:

$$r_1 = 0.45 \pm 0.02, \qquad r_2 = 0.47 \pm 0.02 \ [280],$$

$$r_1 = 0.50 \pm 0.03, \qquad r_2 = 0.45 \pm 0.03 \ [281], \qquad (6.14)$$

determined via the copolymer composition analysis is observed. Hence the different procedures of r_1, r_2 estimation yield the same values within an accuracy of

Table 6.5 Experimental dependences of fractions of triads centered by methyl methacrylate unit and MMA characteristic coefficient (6.15) on monomer feed composition for bulk copolymerization of styrene with methyl methacrylate at T = 60 °C [80]

x_2^0	$P_{22}^{(2)}$	$P_{12}^{(2)}$	$P_{11}^{(2)}$	$K^{(2)}$
0.20	0.01	0.19	0.80	0.99
0.30	0.03	0.28	0.69	1.00
0.40	0.06	0.36	0.58	1.01
0.50	0.10	0.43	0.47	1.00
0.60	0.17	0.48	0.35	1.00
0.70	0.27	0.50	0.23	1.00
0.80	0.42	0.45	0.13	1.01

approximately 10%. The values of the triads, as it may be seen in Table 6.6, calculated knowing the reactivity ratios (6.14) also coincide within the accuracy of the measurement error with their experimental values. It provides convincing evidence on a sufficient reliability of the terminal model and validity of the calculated parameters.

To decide whether the terminal model is applicable to describe the copolymerization of the pair of monomers, it is very convenient to introduce their characteristic coefficients $K^{(s)}$ [18]:

$$K^{(1)} = \sqrt{P_{11}^{(1)}} + \sqrt{P_{22}^{(1)}}, \qquad K^{(2)} = \sqrt{P_{11}^{(2)}} + \sqrt{P_{22}^{(2)}}. \qquad (6.15)$$

The values of these coefficients for the system described within the framework of Mayo-Lewis scheme are to be equal to unity according to relations (6.13) at any monomer feed composition. This condition holds strictly under the copolymerization of styrene with methyl methacrylate (see Table 6.5) while it holds considerably worse under the copolymerization of styrene with acrylonitrile (see Table 6.7). The

Table 6.6 Comparison of theoretical and experimental triad distributions in copolymer of styrene M_1 with methyl methacrylate M_2 prepared in bulk copolymerization (T = 60 °C) at conversions p = 0.03–0.05 [281]

x_2^0	$P_{22}^{(2)}$		$P_{12}^{(2)}$		$P_{11}^{(2)}$	
	Th	Ex	Th	Ex	Th	Ex
0.20	0.01	0.01	0.18	0.21	0.81	0.78
0.33	0.03	0.03	0.30	0.30	0.67	0.67
0.40	0.05	0.05	0.36	0.37	0.59	0.58
0.50	0.10	0.10	0.42	0.40	0.48	0.50
0.60	0.16	0.16	0.48	0.48	0.36	0.36
0.65	0.21	0.21	0.50	0.48	0.29	0.31
0.77	0.36	0.36	0.48	0.44	0.16	0.20
0.80	0.41	0.41	0.46	0.46	0.13	0.13

Table 6.7 Experimental values of composition and microstructure parameters of copolymer of styrene with acrylonitrile prepared in bulk at T = 60 °C. Composition was determined by two techniques: NMR-spectroscopy and elementary nitrogen analysis [215, 282]. Characteristic coefficients were calculated via Eqs. (6.15), (6, 2) according to the experimental data adduced in Refs. [215, 282]

$x_1^0 \cdot 100$	$X_1 \cdot 100$		$P_{11}^{(1)}$	$P_{12}^{(1)}$	$P_{22}^{(1)}$	$K^{(1)}$	$P_{11}^{(2)}$	$P_{12}^{(2)}$	$P_{22}^{(2)}$	$K^{(2)}$
	NMR	NA								
2.1	23.4	—	—	—	—	—	—	—	—	—
5.3	33.3	—	—	—	—	—	—	—	—	—
10.4	40.6	40.0	0.00	0.15	0.85	0.92	0.34	0.55	0.11	0.91
22.1	47.6	48.5	0.00	0.27	0.73	0.85	0.63	0.37	0.00	0.79
31.4	51.0	50.9	0.02	0.34	0.64	0.94	0.69	0.29	0.01	0.93
41.6	54.2	54.8	0.06	0.42	0.52	0.97	0.78	0.22	0.00	0.88
53.0	58.2	59.7	0.07	0.41	0.42	0.91	0.83	0.17	0.00	0.91
63.1	62.7	61.0	0.12	0.56	0.32	0.91	0.88	0.12	0.00	0.94
69.6	64.9	63.8	0.16	0.59	0.25	0.90	0.92	0.08	0.00	0.96
80.2	70.5	69.3	0.30	0.55	0.15	0.94	0.94	0.06	0.00	0.97
88.9	77.2	76.0	0.44	0.50	0.06	0.91	1.00	0.00	0.00	1.00
93.9	82.9	82.7	0.61	0.37	0.02	0.92	1.00	0.00	0.00	1.00

composition and triad fractions of the latter copolymer were thoroughly, analyzed [282] to establish the kinetic scheme of the propagation reactions. This exemplary report convincingly proves that the modern experimental techniques allow one to make a quite reliable choice of the proper copolymerization kinetic model. At first, the authors of Ref. [282] by means of the non-linear least squares procedure [224, 225, 259] determined those sets of the kinetic parameters of the terminal (T), (2.2), penultimate (PU), (2.4), and complex radical (CR) (2.2), (2.6) models that describe the experimental data (see Table 6.7) concerning the dependence of the copolymer composition X_1 on the monomer feed composition x_1^0 to the best. It is worth mentioning that the latter model involves two variants of the selection of the parameters. The first of them (CR-I) assumes the equilibrium constant of the complex formation to be unknown and therefore it was selected together with the reactivity ratios; and the second variant (CR-II) assumes its value to be fixed in accordance with the present literature data. In all four above-mentioned cases the theoretical curves based on the as-selected parameters of the different models perfectly fit the experimentally obtained dependencies of X_1 on x_1^0 within the accuracy of the measurements [282]. This provides additional evidence to the fact that having only the data on the copolymer composition one can not surely decide in favor of any kinetic model. Using the values of the parameters obtained via the copolymer composition analysis, the authors of Ref. [282] through the numerical

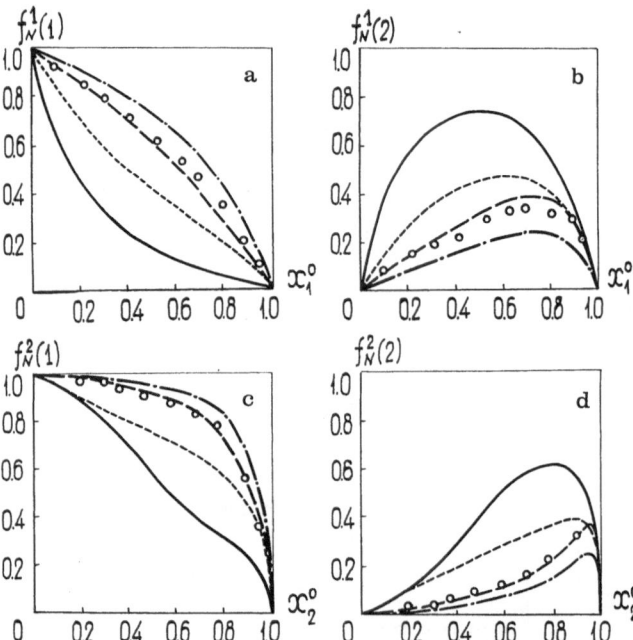

Fig. 16a–d. Fractions of isolated $\bar{M}_i$(**a**), $\bar{M}_2$(**c**) units in copolymer of styrene M_1 with acrylonitrile M_2, and fractions of isolated pairs of these units $\bar{M}_1\bar{M}_1$ (**b**) and $\bar{M}_2\bar{M}_2$ (**d**). Experimental points are compared to the plots calculated via various models: terminal (—∘—), penultimate (— — —), CR-1 (————), CR-2 (--------) [282]

distribution $f_N^{(1)}(n)$ for the run lengths n have calculated the fractions of the isolated units $\bar{M}_1$ of styrene, $f_N^{(1)}(1)$, and isolated pairs of such units, $f_N^{(1)}(2)$. The values of the fractions of the shortest (n = 1, n = 2) runs of acrylonitrile units $\bar{M}_2$, $f_N^{(2)}(1)$ and $f_N^{(2)}(2)$, were calculated in a similar manner. The results of such calculations for all four cases, presented in Fig. 16, reveal a noticeable difference between the microstructure parameters corresponding to the different models. This speaks in favor of the appropriate choice of such parameters for the discrimination of the kinetic models. As it may be seen in Fig. 16, the best agreement between the experimental data and theory is observed when the penultimate model is applied.

Further investigations performed by Australian scientists [283–285] also support this conclusion. Following the general strategy outlined in Sect. 6.1, it was possible by means of the non-linear least squares procedure, to determine correctly [283, 284] the confidence limits for the probabilities of triads. These were calculated within the framework of the different kinetic models taking into account the possible errors arising during the determination of the reactivity ratios. It was also found that the above-mentioned confidence limits corresponding to the penultimate (2.3) and complex-radical (2.5) models practically do not overlap and all the experimental points lie within the first one. This was an additional argument [283] in favor of the reliability of the penultimate model for the description of the bulk copolymerization of styrene and acrylonitrile. The parameters of this model determined independently both via the analyses of the copolymer composition and sequence distribution were found to be rather close to each other [283–285]. This fact also proves the validity of the penultimate model.

An analogous analysis of the solution copolymerization of styrene with acrylonitrile in toluene leads [283] to the same conclusion concerning the choice of the kinetic model as for the bulk copolymerization of these monomers. The applicability of the penultimate model (2.3) was also convincingly proved [283] for the given system, and the estimated values of its four parameters (2.4) (see Table 6.8) were found to be slightly different from the ones obtained in the bulk copolymerization [283]. The experimental values of the fractions of all six triads, determined by means of NMR, in the solution copolymerization products, practically do fit the theoretical plots of the triad fractions vs conversion, which were calculated on the basis of the kinetic parameters presented in Table 6.8.

Table 6.8 Parameters (2.4) of the penultimate model (2.3) describing copolymerization of styrene M_1 with acrylonitrile M_2 in toluene solution at T = 60 °C. The values of reactivity ratios were obtained [283] from the data on copolymer composition (I) and triad distribution (II)

Parameter	I	II
r_1	0.21	0.24
r_1'	0.54	0.55
r_2	0.21	0.22
r_2'	0.08	0.10

Now let us make a short survey of quite different approaches to the problem of identifying the kinetic copolymerization models based on the investigations of the model reactions between the low-molecular compounds only. In this very promising direction it is hard to overestimate the paramount contribution made by Bevington, Tirrell et al. [286–295], whose publications could be divided into two groups.

In the first group [286–293] there were determined the reactivity ratios of the monomer addition to the different low-molecular radicals, the chemical structure of which simulates the structure of the active center at the end of the macroradical. This idea going back to works [286–287] was first used for the polymethacryloni-trile radicals. Then it was employed [288–289] to compare the monomer reactivities towards the 1-phenylethyl radical, modeling the polystyrene radical. The results of the study of the influence of γ-substitutor (similar to the effect of the penultimate units) on the selectivity of styrene or acrylonitrile addition to the simple alkyl radicals were presented [290]. The results of these experiments are in full agreement with the data obtained earlier [282]. It was found that there is a 3.5-fold decrease of the reactivity ratio for the addition of acrylonitrile as compared to styrene under the substitution of γ-phenyl group for γ-cyan one in the attacked radical. The direct observation of such a "penultimate effect" in this simple model system speaks in favor of the penultimate model for the description of the copolymerization of styrene and acrylonitrile. Also similar results for this monomer pair were obtained in the model system consisting of 1-phenylethyl [291] and benzene [293] radicals. As for the former one, the estimated value of the reactivity ratio for styrene relative to acrylonitrile was equal to 0.20 ± 0.02. This value is very close to the corresponding value $r_1 = 0.23$ estimated for the macroradical, the end of which contains two styrene units. It provides more evidence in favor of the penultimate model and reliability of the kinetic parameters of the system under study. However, since its difference from the terminal model (see, for instance, the values of $K^{(i)}$ presented in Table 6.7) is rather insignificant, the latter model also can be employed to calculate the statistical characteristics of styrene-acrylonitrile copolymer in those cases when the high accuracy of the results is not required.

Another series of papers [296–298] should be mentioned, where low-molecular model compounds are used to prove the correctness of the penultimate model of copolymerization. Japanese scientists by means of the ESP-method [297–298] managed to observe a noticeable penultimate effect for the acrylate radical reactivity.

The second set of the works performed by Tirrell et al. [294, 295] concerns the role of the reactions (2.5) of the monomer addition as a binary complex to the growing macroradical end. The latter is simulated by the corresponding low-molecular compound, adding the monomers one by one as well as in pairs as complexes, which transforms finally into a non-active product after the interaction with the radical trap. The analysis of these products reveals that in the system under study, the reactions (2.5) proceed quite insignificantly and then one can ignore them deciding on the proper kinetic model of copolymerization [294, 295].

It should be pointed out that the computer algorithms and programs [299, 300, 197, 301, 99, 302, 103, 127, 86, 102, 93], developed to calculate the copolymer statistical characteristics within the frameworks of the different models and to estimate their parameters from the experimental data, are considered to be rather useful to solve the problems of such a choice. The leading scientists in the field of radical polymerization also emphasized the utmost importance of such computer approaches in delivered at a recent representative conference in Genoa [265, 283, 303].

O'Driscoll et al. [304] also demonstrated how carefully the problems of copolymerization model discrimination should be considered. They noticed that although the correct treatment of the experimental data on the copolymer composition requires the non-linear least squares procedure, the linear approaches remain still popular among researchers studying radical copolymerization. Within the framework of the latter approaches the objective criterion for the judgement whether the terminal model (2.1) can be employed is the straightening of the experimental plots of X versus x^0 in the corresponding coordinates, the choice of which depends on the used approach (see Sect. 6.2). For instance, according to the Kelen-Tudos procedure [217] such coordinates are the variables η and ξ of Eq. (6.7), and therefore the curvilinear character of the plot of η vs ξ manifests the inapplicability of the Mayo-Lewis model (2.1) for a given copolymerization system. The authors of Ref. [304] expressed doubts concerning such a conclusion pointing out that the curvilinear character of the plot of η vs ξ may arise from the incorrect account of the errors in the framework of linear least squares procedure. To confirm this statement they performed a computer experiment generating the values of the copolymer composition for various reactivity ratios and inserting random errors into them. The conclusion [304] was that the linear analysis of the copolymer composition data can not by itself be a reliable tool of model discrimination. For instance, the noticeable curvilinearity of the η vs ξ plot can ensue from a random analytical error of about 3% in the system where $r_1 = 0.1$ and $r_2 = 10.0$.

The selection of the proper kinetic model to describe the copolymerization at high conversion acquires ever greater importance since industrial processes are carried out precisely under such regimes. This problem requires a thorough experimental examination and the correct treatment of the obtained results. The excellent work of Ref. [305] performed by O'Driscoll et al. is an illustrative example. They studied the drift of the composition in the course of the bulk copolymerization of styrene with methyl methacrylate up to a conversion of $p = 0.7$. The data of the earlier studies [306–308] on this system brought their authors to contradictionary conclusions both in favor [306] and against [307–308] the terminal model. To ascertain the truth, the authors of Ref. [305] by means of spectroscopic methods measured the drift of the monomer feed composition $x(p)$ (by NMR) and the drift of the average copolymer composition, $\langle X(p) \rangle$ (by NMR and UV). The closeness of the results obtained via these three independent techniques furnishes evidence that they are reliable enough. The further treatment of these results through the "errors in variables" procedure, which now is the most common variant of the non-linear least squares analysis [239, 241, 247], results in the well-grounded conclusion [305] that some previous copolymer analyses [307, 308] were erroneous and the Mayo-Lewis model in fact is very well applicable to the system under

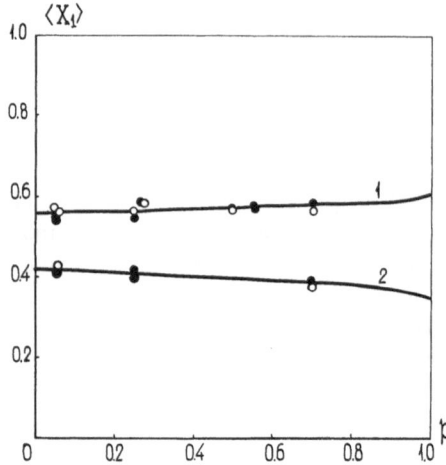

Fig. 17. Dependencies of average composition $\langle X_1 \rangle$ of copolymer of styrene with methyl methacrylate on the weight conversion of reacted monomers p' at the initial compositions $x_1^0 = 0.6$ (*1*) and $x_1^0 = 0.35$ (*2*). The curves are calculated; experimental data, obtained by means of NMR and UV spectroscopy, are depicted by *dark* and *open circles*, respectively [305]

consideration. Moreover, the values of the reactivity ratios, $r_1 = 0.50$ and $r_2 = 0.46$, obtained in Ref. [305] practically do coincide with those values, $r_1 = 0.50 \pm 0.03$ and $r_2 = 0.47 \pm 0.03$, that were presented earlier [227] on the basis of the analysis of a number of literature data on the copolymerization of styrene and methyl methacrylate. The drift of the average copolymer composition with conversion, calculated according to these values of r_1 and r_2, fits the experimental data rather well (see Fig. 17). The applicability of the terminal model for the calculation of the copolymerization at high conversions for the same system was convincingly established [309] via the copolymer triad distribution determined by means of NMR spectroscopy.

Unfortunately, as far as the author knows, there are only a few publications where the problem of the validity of this or that model over a wide range of conversions and initial monomer feed compositions was discussed carefully enough. Here one might mention the works listed as Refs. [310, 201] on the bulk copolymerization of styrene and heptyl acrylate, where the adequacy of the terminal model was undoubtedly proved, and its parameters $r_1 = 0.87$ and $r_2 = 0.27$ were estimated. Really, the calculated copolymer composition and monomer feed composition drift with conversion are in full agreement with both NMR (Table 6.9) and UV (Fig. 18) data.

So trustworthy arguments furnished by the authors of Ref. [283] testify that the copolymerization of styrene and acrylonitrile in toluene solution can be fairly

Table 6.9 Comparison of the values of initial copolymer composition X_{st} calculated at $r_{st} = 0.87$, $r_{ha} = 0.27$ (X_{st}^{th}) and those obtained from the analysis of ^{1}H-NMR spectra (X_{st}^{ex}) at various monomer feed compositions x_{st}^0 [310]

x_{st}^0	0.1	0.2	0.3	0.4	0.5	0.6	0.7	0.8	0.9
X_{st}^{th}	0.24	0.37	0.45	0.53	0.59	0.66	0.73	0.81	0.89
X_{st}^{ex}	0.27	0.38	0.51	0.57	0.59	0.66	0.75	0.80	0.91

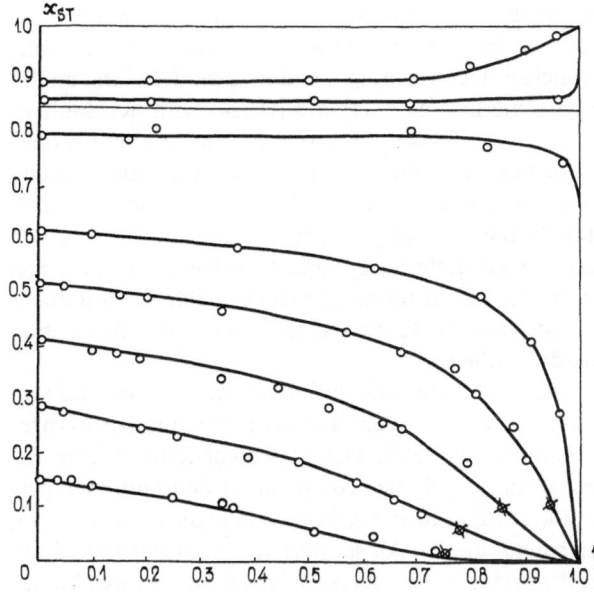

Fig. 18. Comparison of the experimental points and theoretical plots of the monomer feed composition x_{st} on conversion p at different initial values of x_{st}^0. *Crosses* denote the moments of the transparency loss in the reaction system [201]

described by the penultimate model over a wide range of conversions. This was strongly confirmed by a very high agreement of the experimental values of the triad fraction with those calculated on the basis of the kinetic parameters listed in Table 6.8. Such an agreement was observed over the whole studied range of the initial monomer feed compositions, namely, $0.1 < x_1^0 < 0.8$. Two typical examples that illustrate the above statements are presented in Fig. 19.

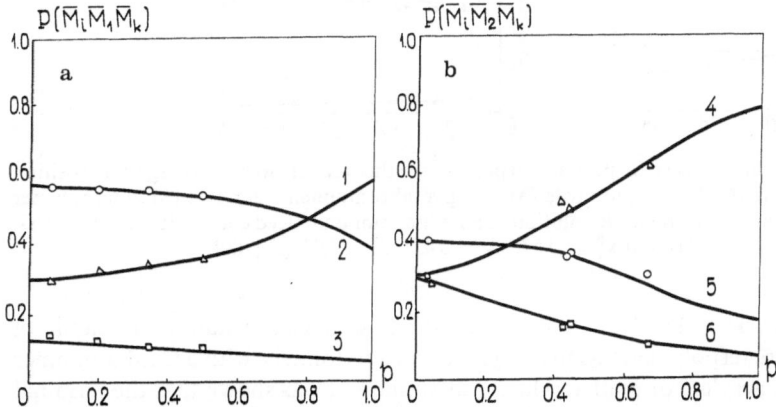

Fig. 19a, b. Comparison of experimental points and theoretical plots calculated within the penultimate model depicting the dependence of fractions of triads $P(\bar{M}_1\bar{M}_1\bar{M}_1)$ *(1)*, $P(\bar{M}_1\bar{M}_1\bar{M}_2)$ *(2)*, $P(\bar{M}_2\bar{M}_1\bar{M}_2)$ *(3)*; $P(\bar{M}_2\bar{M}_2\bar{M}_2)$ *(4)*, $P(\bar{M}_1\bar{M}_2\bar{M}_2)$ *(5)*, $P(\bar{M}_1\bar{M}_2\bar{M}_1)$ *(6)* on conversion for the copolymer of styrene (M_1) and acrylonitrile (M_2) prepared in toluene solution at an initial monomer feed composition $x_1^0 = 0.81$ **(a)** and $x_1^0 = 0.125$ **(b)** [283]

Thus, through the body of the mentioned experimental evidence obtained via different methods that characterize the composition and structure of macromolecules one arrives at a simple conclusion concerning the kinetic model of the binary copolymerization of styrene with methyl methacrylate (I) and with acrylonitrile (II). The former of these systems is obviously described by the terminal model, and the latter one by the penultimate model. However, the latter system characteristics in those cases when high accuracy of the results is not required, may be calculated within the framework of the Mayo-Lewis model. Such a simplified approach was found to be quite acceptable to solve many practical problems. One should note that the trivial terminal model is able to describe a vast majority (at least, 90% according to Harwood [303]) of copolymerization systems which have been already studied.

The applicability of this model for the calculation of the composition of terpolymers produced at low conversions was demonstrated for a number of three component systems (see, for example, [6, 132]). This statement follows from the good agreement of the values of the terpolymer composition determined experimentally with those calculated according to the Alfrey-Goldfinger equations [45] within the framework of the terminal model. All six parameters of this model are considered to be known from the preliminary analysis of the data on three binary systems selected from the given three monomers.

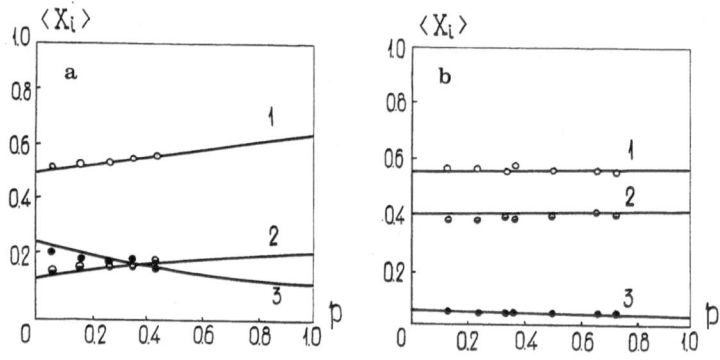

Fig. 20a, b. Comparison of experimental points with theoretical plots of average composition $\langle X_i \rangle$ of [styrene (M_1) + acrylonitrile (M_2) + pentabromophenyl acrylate (M_3)] terpolymer prepared in dimethyl formamide solution at an initial monomer feed compositions $x_1^0 = 0.68$; $x_2^0 = 0.21$; $x_3^0 = 0.11$ (**a**) and $x_1^0 = 0.57$; $x_2^0 = 0.41$; $x_3^0 = 0.02$ (**b**) [200]

Several works [311–313, 200] are devoted to a circumstantial research of systems where terpolymerization of styrene and acrylonitrile with a third monomer (brominated acrylate or methacrylate) was studied. It was shown that the terminal model can be employed to describe such systems. One can see this from Table 6.10 and Figs. 20 and 21, where the conversion drifts of the average copolymer composition are presented.

A conclusion to be drawn from the present section is that the accuracy of both the modern experimental techniques and the mathematical treatment of the

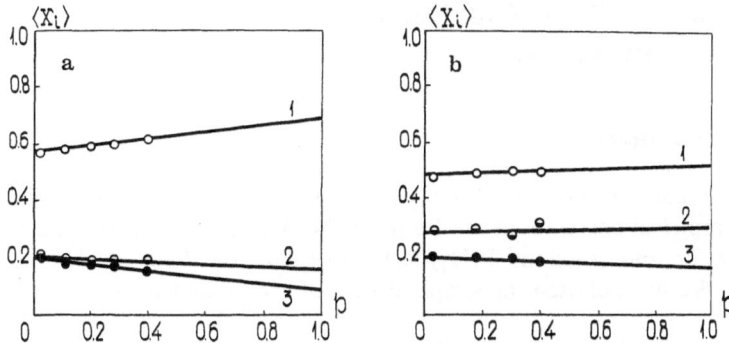

Fig. 21a, b. Plots similar to those presented in Fig. 20 but for terpolymer of [styrene (M_1) + acrylonitrile (M_2) + 2, 4, 6 — tribromphenyl methacrylate (M_3)] at $x_1^0 = 0.71$; $x_2^0 = 0.18$; $x_3^0 = 0.11$ (**a**) and $x_1^0 = 0.51$; $x_2^0 = 0.32$; $x_3^0 = 0.17$ (**b**) [313]

Table 6.10 Comparison of experimental (X_i^{ex}) and theoretical (X_i^{th}) values of compositions of terpolymers (styrene (M_1) + acrylonitrile (M_2) + bromine-containing monomer (M_3)) prepared in dimethylformamide at low conversions ($p < 0.1$) and at various monomer feed compositions (x_i^0) [312, 313]. 2,4,6-Tribromine phenyl ether of acrylic (I) [312] and methacrylic (II) [313] acids were used as M_3

System	x_1^0	x_2^0	x_3^0	X_1^{ex}	X_2^{ex}	X_3^{ex}	X_1^{th}	X_2^{th}	X_3^{th}
I 1	0.15	0.59	0.26	0.32	0.44	0.24	0.32	0.46	0.22
I 2	0.66	0.19	0.15	0.50	0.27	0.23	0.54	0.24	0.22
I 3	0.71	0.18	0.11	0.55	0.25	0.20	0.56	0.26	0.18
I 4	0.62	0.35	0.03	0.50	0.45	0.05	0.52	0.44	0.04
I 5	0.41	0.28	0.31	0.46	0.23	0.31	0.47	0.24	0.29
I 6	0.25	0.41	0.34	0.41	0.29	0.30	0.41	0.30	0.29
I 7	0.33	0.65	0.02	0.41	0.56	0.03	0.40	0.58	0.02
I 8	0.57	0.35	0.08	0.50	0.38	0.12	0.51	0.39	0.10
I 9	0.49	0.10	0.41	0.46	0.11	0.43	0.50	0.09	0.41
I 10	0.02	0.60	0.38	0.13	0.50	0.37	0.10	0.53	0.37
II 1	0.71	0.18	0.11	0.58	0.21	0.21	0.59	0.21	0.20
II 2	0.52	0.40	0.08	0.48	0.38	0.14	0.51	0.36	0.13
II 3	0.20	0.50	0.30	0.33	0.31	0.36	0.31	0.33	0.36
II 4	0.09	0.78	0.13	0.25	0.52	0.23	0.22	0.55	0.23
II 5	0.56	0.26	0.18	0.51	0.22	0.27	0.51	0.23	0.26
II 6	0.34	0.32	0.33	0.44	0.20	0.36	0.40	0.23	0.37
II 7	0.39	0.49	0.12	0.44	0.38	0.18	0.45	0.38	0.17
II 8	0.39	0.39	0.22	0.41	0.30	0.29	0.43	0.29	0.28
II 9	0.52	0.24	0.24	0.46	0.20	0.34	0.49	0.20	0.31
II 10	0.52	0.19	0.29	0.48	0.15	0.37	0.49	0.16	0.35

experimental data obtained by their means, allows to safely discriminate the kinetic models of copolymerization for concrete systems and to estimate reliably the values of their parameters.

7 The Effect of Synthesis Conditions on Copolymer Properties

7.1 Binary Copolymers

Heat resistance is regarded to be one of the well-studied properties of copolymers. Many different empirical correlations are known [314–316] between glass transition temperature T_g and fractions $P(U_2) \equiv P(\bar{M}_i\bar{M}_j)$ of the diads $(\bar{M}_i\bar{M}_j)$ in macromolecules. We have chosen the simplest form of such a dependence:

$$T_g = T_{11}P(\bar{M}_1\bar{M}_1) + T_{22}P(\bar{M}_2\bar{M}_2) + T_{12}P(\bar{M}_1\bar{M}_2) \qquad (7.1)$$

where T_{11}, T_{22}, and T_{12} are glass transition temperatures of homopolymers and of alternating copolymer, respectively. The latter value is known for a series of systems [317] since a procedure of preparing regularly alternating copolymers has already been developed. When synthesis of such copolymers is not possible, the value of T_{12} can be determined [318] from the slope of the plot of $Y = T_g - T_{11}P(\bar{M}_1\bar{M}_1) - T_{22}P(\bar{M}_2\bar{M}_2)$ versus $P(\bar{M}_1\bar{M}_2)$ which is corresponded to Eq. (7.1). When T_{11}, T_{22}, and T_{12} are known, copolymer glass transition temperature can be calculated within the framework of any kinetic model, describing copolymer formation, according to Eq. (7.1) or some other similar equation. Essentially, additional adjusting parameters are not needed for such a calculation. The comparison of experimental data on T_g with the calculated values usually results in their agreement [18] (see Fig. 22).

A number of important physicochemical properties of copolymers depends to a great extent on their composition inhomogeneity. This has been demonstrated by Nielsen [319], who studied the influence of the composition distribution of copolymers of styrene with methyl acrylate (1), and vinyl chloride with methyl acryate (2) on some of their performance characteristics. He compared a series of mechanical properties of two copolymer samples which were characterized by the same average composition but by different compositional inhomogeneity. It was found that the more homogeneous the sample was, the stronger the dependence of shear modulus on temperature was pronounced. Moreover, in this case a narrower maximum of the plot of the logarithm of decrement dumping versus temperature was observed. It is worth mentioning that the differences between these two samples (one of them was prepared at low conversion, and the other at higher conversion) were more pronounced for copolymer (1). This might be explained by the more essential difference in composition inhomogeneity of the pair of samples which turns out to be greater for the second copolymer than for the first one. The reason for such a difference is connected with the distinction of the reactivity ratios as it can be easily estimated by elementary calculations.

Some interesting results have been obtained by Russian scientists [320, 321] who studied the influence of composition inhomogeneity on some service properties and supermolecular structures of copolymers. Two samples of copolymers of butyl acrylate with methacrylic acid were synthesized which had a similar average

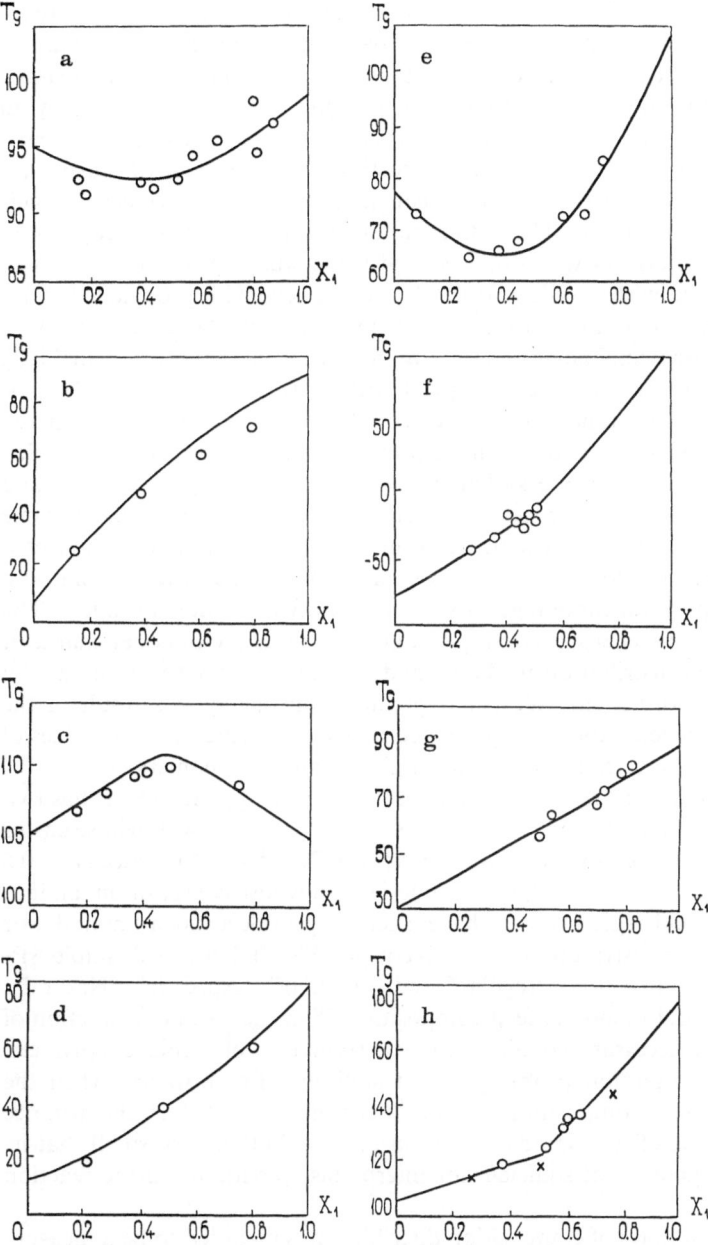

Fig. 22a–h. Glass transition temperature versus composition of copolymers: methyl metha-crylate + styrene (**a**); styrene + methyl acrylate (**b**); acrylonitrile + styrene (**c**); vinyl chloride + methyl acrylate (**d**); methyl methacrylate + vinyl chloride (**e**); acrylonitrile + buta-diene (**f**); acrylonitrile + vinyl acetate (**g**); α-methyl styrene + acrylonitrile (**h**). Experimental points obtained at low conversions from various publications, are compared to the theoretical plots calculated according to Eqs. (7.1) within the framework of the terminal model [18]

compositon $X_1 \approx 0.3$ but different composition inhomogeneity. This difference resulted from their preparation: the first sample of the copolymer was obtained at a conversion of 5% and the second one at 80%. It was found that the breaking strength of the more homogeneous sample was higher than the other one by an order of magnitude. The authors of Ref. [321] concluded that there is a principal possibility to improve the mechanical properties of copolymers, (in particular, when they consist of polar and non-polar monomer units) decreasing their composition inhomogeneity. The influence of this factor on the copolymer performance characteristics was also observed while studying the copolymers of methyl methacrylate with acrylic acid [322], and ethylene with propylene [323].

The composition inhomogeneity had a strong effect on the compatibility of copolymers as it was convincingly shown by Skeist in his pioneering work [12]. Studying copolymerization of styrene (M_1) with methyl methacrylate (M_2) he noticed that some important properties of copolymerization products, prepared at complete conversions, were quite different for rich and poor in styrene samples. It was observed that when one system ($x_1^0 = 0.8$; $x_2^0 = 0.2$) was still light and transparent during the whole process, another system ($x_1^0 = 0.2$; $x_2^0 = 0.8$) became turbid and the copolymer obtained in the latter was more brittle. Skeist connected the phase separation followed by a transparency loss in the latter system with rather high composition inhomogeneity of its copolymerization products. The reason of this inhomogeneity according to Skeist [12] was the bimodal character of the composition distribution in the second system contrary to the unimodal distribution in the other one. We could predict immediately this result in the framework of the general approach presented in Sect. 5.6 using only the values of reactivity ratios ($r_1 = 0.75$; $r_2 = 0.18$) for a given pair of monomers.

Actually, this system corresponds to a phase portrait of type II (5.8). It has two stable stationary points (SPs); ($x_1^* = 1$; $x_2^* = 0$) and ($x_1^* = 0$; $x_2^* = 1$) whose basins of attraction are separated by the azeotrope ($x_1^* = 0.77$; $x_2^* = 0.23$). Since $r_1 > 0.5$ and $r_2 < 0.5$, products of complete copolymerization whose composition fall into the first or the second region, should have, according to Sect. 5.6, a unimodal or bimodal composition distribution, respectively (see Fig. 4). Let us call stable SPs with such basins of attraction "regular" and "singular", respectively. Naturally the locus of the initial monomer feed composition $\bar{x}^0$ in the basin of attraction of a singular SP is a necessary condition for the formation of turbid copolymers. Nevertheless, this condition probably is not sufficient. For instance, when the starting monomer feed composition $\bar{x}^0$ is close enough to this SP $\bar{x}^*$, the distance between the maxima of the bimodal composition distribution is so small that its inhomogeneity degree is not sufficient for microphase separation in the reaction system.

Since the dispersion σ^2 of composition distribution is regarded to be a measure of such an inhomogeneity, it is straightforward to suppose that when σ^2 exceeds its critical value σ_{cr}^2, the polymer becomes turbid. So, the inequality $\sigma^2 > \sigma_{cr}^2$ can be considered as a sufficient condition for the copolymer to lose its transparency. The value of σ_{cr}^2 can be approximately considered as a universal parameter of the system which depends only on temperature [134, 6]. To verify this assumption, the conversions p^+, at which the system becomes turbid, were experimentally

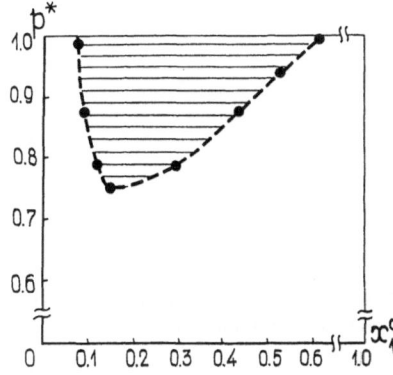

Fig. 23. The boundary between turbid *(shaded area)* and transparent regions of copolymer of styrene with heptyl acrylate [310]

determined in the course of bulk copolymerization of styrene and heptyl acrylate at various initial monomer feed compositions. In some boundary points of the turbidity region, shown in Fig. 23, there were calculated the values of the average copolymer composition $\langle x_{st} \rangle$ and dispersion $\sigma^2(p^+)$ of its composition distribution. These data are presented in Table 7.1. One can see that despite a certain difference in copolymer composition and conversion p^+ (and consequently viscosity), at which transparency disappears, microseparation occurs at practically the same value of $\sigma^2(p^+) = (5.5 \pm 1) \times 10^{-3}$. Therefore this value is a critical dispersion σ^2_{cr} for the copolymer of styrene (ST) and heptyl acrylate (HA). The slight difference in values of $\sigma^2(p^+)$ arises due to the experimental accuracy ($\pm 1\%$) of p^+ determination as well as to the possible errors in r_{ST} and r_{HA} estimates. This difference is small enough in comparison with the range of dispersion alteration, $0 < \sigma^2(p) < \sigma^2(1)$, where the value of $\sigma^2(1)$, for example at $X^0_{ST} = 0.41$ is equal to 20×10^{-3}.

In order to elucidate the effect of temperature, the authors of Refs. [310, 210] determined experimentally the boundary points $x^0_{ST} = 0.08$ and $x^0_{ST} = 0.65$ of the transparency region for the (ST + HA) system at complete conversion, $p = 1$, when in the course of synthesis the temperature was increased in a given way from 28 °C to 78 °C. Despite a noticeable difference between such a regime and the isothermal one (see Fig. 24), it was found that the regions, in which at $p = 1$ turbid copolymers were formed, practically coincide. The same could be said about the calculated values of dispersion $\sigma^2(1)$ at the boundary points of the mentioned regions. This might be associated with a rather weak dependence of the reactivity ratios on temperature. A similar practical independence of the turbidity region

Table 7.1 Statistical characteristics of the composition distribution at boundary points of the turbidity region [310, 201]

x^0_{st}	0.08	0.09	0.12	0.15	0.29	0.41	0.52	0.62
p^+	0.99	0.85	0.79	0.75	0.78	0.87	0.95	1.0
$\langle X_{st} \rangle$	0.08	0.11	0.15	0.20	0.36	0.46	0.54	0.62
$\sigma^2(p^+) \, 10^3$	4.6	4.9	6.2	6.6	6.0	5.3	5.1	6.4

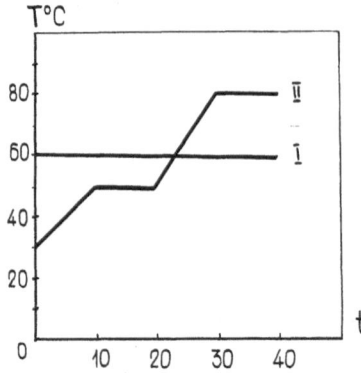

Fig. 24. Isothermal (I) and temperature-control-led (II) regimes of copolymerization [310]

boundaries on temperature was ascertained [310, 201] for the copolymerization of styrene with methyl acrylate (MA). It was found that in the (ST + MA) system transparent copolymers were formed outside the regions $0.03 < x_{ST}^0 < 0.61$ (Regime I) and $0.02 < x_{ST}^0 < 0.63$ (Regime II) at complete conversion. The above-mentioned examples allow one to predict the transparency of copolymers prepared in Regime II as well as in Regime I. Dispersion $\sigma^2(1) = 5.6 \times 10^{-3}$ calculated according to Eq. (5.7) with the account of $r_{ST} = 0.75$, $r_{MA} = 0.18$, at the boundary point $x_{ST}^0 = 0.63$ of the (ST + MA) system, practically coincides with $\sigma_{cr}^2 = 5.5 \times 10^{-3}$. This apparently shows that for the synthesis of various styrene-acrylic copolymers the values of critical dispersions are very close to each other.

In order to prove that the high composition inhomogeneity was the only reason for microphase separation both (ST + HA) and (ST + MA) copolymers, some of their samples with the same compositions prepared, however, at low conversions (p < 0.10) of monomers with their subsequent distillation were studied [310, 210]. All these copolymers whose dispersion is by one or two orders of magnitude less than that of the products of complete conversion (p = 1), turn out to be clear.

7.2 Possibilities of Prediction of Some Properties of Multicomponent Copolymers

The advantage of good theory is that it permits an explorer to obtain important results via purposeful search avoiding occasional tedious selection of different variants. In the latter case the probability of optional choice of the conditions of multicomponent copolymer synthesis is almost the same as in lottery. So, only a minor part of the variants tested in laboratories can be applied in practice. The idea allowing one to predict effectively the properties of multicomponent copolymers implies the possibility to avoid much tedious routine experimentation over the entire range of compositions of multicomponent systems, using experimental data on binary systems only [6, 18]. Employing these data as starting information in a number of cases one can predict via mathematical modeling particular properties of multicomponent copolymers in a simple way.

For example, to calculate the glass transition temperature, one should use the following equation [18]:

$$T_g = \sum_{j=1}^{m} \sum_{i=1}^{j} T_{ij} \langle P(\bar{M}_i \bar{M}_j) \rangle \qquad (7.2)$$

which is known to be a generalization of Eq. (7.1) in the case of m-component copolymer and arbitrary conversions. Glass transition temperatures of homopolymers T_{ii} and alternating copolymers T_{ij} ($i \neq j$) are the parameters of the above equation. They can be presented as a symmetric matrix with $m(m + 1)/2$ independent elements. To calculate T_g according Eq. (7.2) one should in addition to the mentioned matrix use only the matrix of reactivity ratios with $m(m - 1)$ elements of r_{ij} ($i \neq j$) determined in binary systems under the copolymerization of monomers ($M_i + M_j$). Using this matrix, one can easily via expressions (5.1) and (5.2) compute the values of the average fractions $\langle P(\bar{M}_i \bar{M}_j) \rangle$ of dyads for m-component copolymer at arbitrary monomer feed composition and conversion.

Along with heat resistance the transparency of copolymers is regarded to be one of the most important properties. Therefore it is worth mentioning the important work by Slocombe [128], who studied experimentally the dependence of transparency of a number of terpolymers, prepared at conversions close to complete, on the monomer feed composition. Having investigated 44 such three-component systems, each of which had no less than two binary azeotropes, Slocombe managed to find out some interesting empirical regularities.

It was discovered [128], that for a majority of studied systems, the clear copolymers were produced when the monomer feed composition lay in the vicinity of the straight line (called "azeotropic line") connecting a pair of binary azeotropes located on sides of the Gibbs-Roozeboom triangle. Such systems, as, for example, shown in Fig. 25a, were called "regular" to differentiate them from "irregular" ones that did not follow this rule. Slocombe, while analyzing experimental data, noticed that in many regular systems, transparency loss of some terpolymers arose because of the slight deviations of initial monomer feed compositons from the vicinity of the azeotropic line.

Besides studying some regular systems (for example, shown in Fig. 25a) Slocombe discovered one more interesting experimental peculiarity. Terpolymers, the compositions of which were quite different but lay in the vicinity of the azeotropic line, were found to be thermodynamically compatible. However, by mixing terpolymers with much less difference in their compositions but which lay at the opposite sides of the azeotropic line, one would obtain turbid incompatible blends. The developed theorey [6, 134] (see Sect. 5.6) provides an opportunity to establish whether a system with given matrix of reactivity ratios is regular and allows one to explain the empirical regularities discovered by Slocombe.

In the literature we found a complete set of reactivity ratios only for eleven systems (nine regular and two irregular) studied by Slocombe [128]. Further numerical calculations showed [6, 134] that boundary trajectories (separatrices) lay quite close to the azeotropic lines. It also turned out that for all nine regular systems the separatrices entered a regular binary azeotrope $\bar{x}^*$, while in both

82

Semion I. Kuchanov

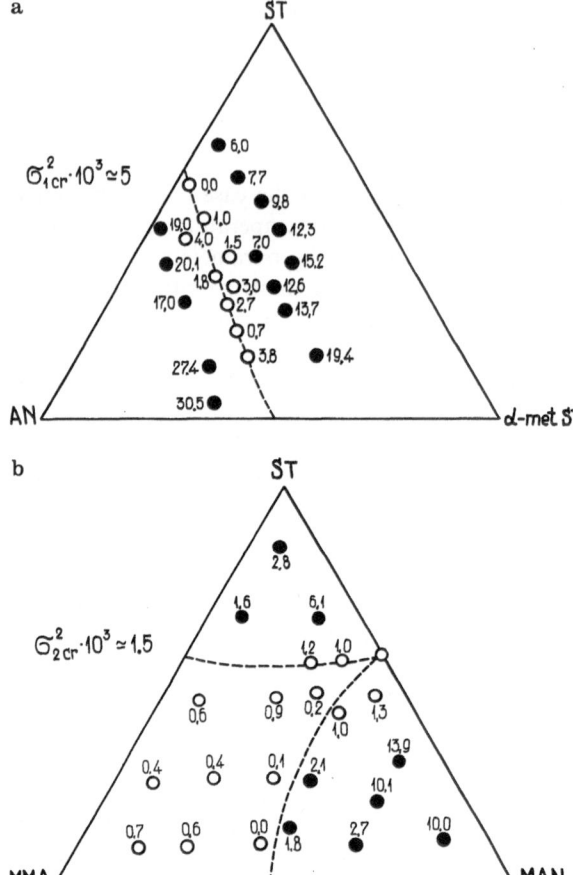

Fig. 25a, b. Correlation between transparency and calculated [6, 201] dispersion $\sigma_i^2(1) \times 10^3$ of composition distribution of terpolymer molecules, prepared at complete conversion, for the reference i-th monomer in the system: (acrylonitrile + α-methylstyrene + styrene) (**a**) and (methyl methacrylate + methacrylonitrile + styrene (**b**). Open and dark circles denote transparent and turbid terpolymerization products in the experiments by Slocombe [128]

irregular systems they entered a singular binary azeotrope $\bar{x}^*$. Hence, according to general theory developed in Sect. 5.6, the products of terpolymerization with compositions close to the azeotropic line will be much more compositionally homogeneous in the first than in the second case. This might serve as an explanation for the differences in their transparency discovered by Slocombe.

The same situation takes place in the system shown in Fig. 25a, which according to general classification (see Fig. 7) refers to the kind 10. Both stable stationary points (SPs) located at the apexes of the triangle base are singular in contrast to regular boundary binary azeotrope. Hence according to the general theory developed in Sect. 5.6, one could expect compositionally inhomogeneous terpolymers produced within the whole Gibbs-Roozeboom triangle except the azeotropic line, practically coinciding with the separatrix (see Fig. 25a). In fact, experimental data presented by Slocombe in Fig. 25a have one-to-one correspondence with such qualitative theoretical conclusions.

In order to make these conclusions quantitative, it is natural to correlate the transparency loss of the system with terpolymer composition inhomogeneity towards acrylonitrile. The choice of this monomer M_1 as a determining one is due to its noticeable difference (for example, by polarity) from the other two monomers: styrene and α-methyl styrene which are quite similar. This approximation seems to be reasonable. It follows from Fig. 25a, where one could see the calculated dispersions $\sigma^2(1)$ of acrylonitrile composition distribution of the products at complete conversion $(p = 1)$ for those values of compositions of terpolymers, the transparency of which have been studied by Slocombe. Actually, clear products are formed if dispersion $\sigma^2(1)$ does not exceed some critial value σ^2_{1cr}, wchih lies between 4×10^{-3} and 6×10^{-3}. In the opposite case $\sigma^2_1(1) > \sigma^2_{1cr}$, as it may be seen from Fig. 25a, only turbid terpolymers are formed.

The above-mentioned regularity was found to be useful [201] to predict the transparency of terpolymer mixtures (acrylonitrile + styrene + α-methyl styrene) of various compositions, which had been studied by Slocombe [128]. He studied the turbidity of sixteen binary and one tetranary blends, prepared via mixing of eight samples corresponding to the particular points of the Gibbs-Roozeboom triangle taken in various combinations. The calculations showed that dispersion σ^2_1 of all transparent blends was less than its critical value σ^2_{1cr}, meanwhile for all turbid mixtures $\sigma^2_1 > \sigma^2_{1cr}$. To calulate the dispersion of i-th monomer composition distribution $f(\xi_i)$ of n-component mixture the following equation was used [201]:

$$\sigma^2_i = \sum_{v=1}^{n} \alpha_v \sigma^2_{vi} + \sum_{v<\mu}^{n} \alpha_v \alpha_\mu (x^0_{vi} - x^0_{\mu i})^2 \qquad (7.3)$$

where $X_{vi} = x^0_{vi}$ and $\sigma^2_{vi}(v = 1, ..., n)$ are average value and dispersion of M_i monomer distribution of $f_v(\xi_i)$ in the copolymer sample which is the v-component of the mixture, the fraction of which is α_v.

The calculation of dispesion σ^2_1 performed through Eq. (7.3) for a number of mixtures of the terpolymer (acrylonitrile + styrene + α-methyl styrene) samples are consistent with empirical rules by Slocombe concerning the compatibility of regular polymers of various compositions. One can easily understand the reason for such an accordance. So terpolymerization products of composition $\bar{x}^0$ lying even at opposite ends of the azeotropic line, contain practically the same fraction of acrylonitrile. Therefore, the absolute value of the contribution to dispersion σ^2_1(7.3) from the second sum is rather small. The same could be said about the first sum in Eq. (7.3) since all dispersions σ^2_{v1} are small due to the regularity of the considered system. Hence, dispersion σ^2_1 (7.3) of composition distribution $f(\xi_1)$ of a mixture of terpolymers, the compositions of which lie on the azeotropic line, will be rather negligible. This stipulates transparency of such blends. Three-component systems with three binary azeotropes have also been studied [128] and for one of them the experimental data on the transparency of the copolymerization products over the entire range of their compositions were adduced. The data presented in Fig. 25b admit a simple interpretation within the framework of the general approach (see Sect. 5.6). Hence knowing the values of reactivity ratios one can easily conclude that: (a) a given three-component system refers to kind 14 according to the classification presented in Fig. 7; (b) among three stable stationary points (SPs) at the apexes of the triangle only one located in its left corner is regular. This is just the reason why in its basin transparent copolymers are formed, while the copolymers formed in the basins of two other SPs are turbid. Since the products of binary copolymerization of styrene and methyl methacrylate are transparent over the whole range of their compositions, it is quite natural to choose methacrylnitrile as a reference monomer for a given three-component system. The comparison of the experimental data with the calculated values of dispersion $\sigma^2(1)$ presented in Fig. 25b proves the correctness of the simple condition $\sigma_2^2(1) > \sigma_{2cr}^2$ of terpolymer turbidity. The value of σ_{2cr}^2 lies between 1.3×10^{-3} and 1.6×10^{-3} for a given system.

Thus the above results, seemingly, gave evidence that the developed theoretical approach could effectively explain a number of experimental regularities concerning the transparency of several previously studied terpolymers. This allows one to look forward optimistically to the success of such an approach in predicting a given property in the stage of the synthesis while developing processes of manufacturing of some new multicomponent copolymers. The investigations we performed [310, 201] for copolymerization of styrene with acrylic and methacrylic monomers is a strong argument in favor of this approach.

First let us demonstrate the possibilities of predicting transparency and heat resistance of (styrene + methylacrylate + heptyl acrylate) terpolymerization of the products prepared at complete conversion of monomers. The elements r_{ij} of the matrix of reactivity ratios:

$$\begin{bmatrix} 1 & 0.75 & 0.87 \\ 0.18 & 1 & 1.24 \\ 0.27 & 0.55 & 1 \end{bmatrix} \tag{7.4}$$

calculated [310, 201] on the basis of experimental data on binary systems (styrene + methyl acrylate), (styrene + heptyl acrylate) and (methyl acrylate + heptyl acrylate) allow one to calculate the dispersion $\sigma_1^2(1)$ of the composition distribution with styrene as a reference monomer for terpolymer samples of

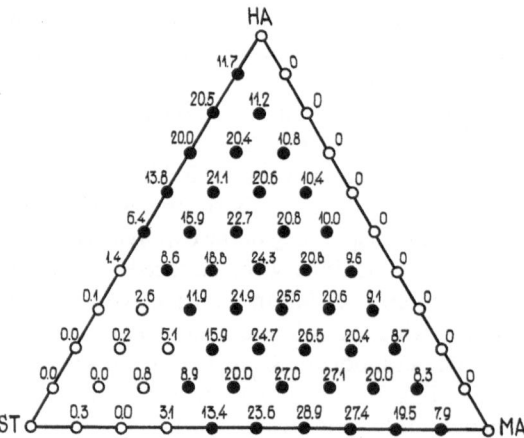

Fig. 26. The Gibbs-Roozeboom triangle for the weight composition of (styrene + methyl acrylate + heptyl acrylate) terpolymer, prepared under Regime II (see Fig. 24) at complete conversion. Open and dark circles denote transparent and turbid terpolymer, respectively, and figures above the circles denote the theoretical value of dispersion $\sigma_1^2(1) \times 10^3$ [310, 201]

arbitrary composition. The results of such a procedure are shown in Fig. 26 for the whole Gibbs-Roozeboom triangle. According to the theory, if for some compositions the calculated dispersion $\sigma_1^2(1)$ exceeds its critical value $\sigma_{1cr}^2 \approx 5.5 \times 10^{-3}$, the turbid copolymer would be formed. Contrarily, those points of the triangle where $\sigma_1^2(1) < \sigma_{1cr}^2$ should be located within the transparency region. As it might be seen from Fig. 26, these regularities are observed for terpolymers of all studied compositions, with no exceptions.

It is worth mentioning that the qualitative pattern of the regions of transparency and turbidity in the Gibbs-Roozeboom triangle (Fig. 26) can be predicted even before computing the dispersions σ_1^2. Thus, elementary theoretical analysis shows that a given system with the matrix of reactivity ratios (7.4) refers to kind 12 in Fig. 7. This system has two stable stationary points (SPs) located at the left and upper triangle apexes. Each of these SPs have its own basin of attraction. In the smaller basin, where the system is rich in styrene, the theory predicts the formation of transparent terpolymers, since SP in the left triangle apex is regular. In the other part of the triangle which is the basin of attraction of the other stable SP due to its singularity, one should expect turbid terpolymers to be formed. These qualitative forecasts are real, as it may be seen in Fig. 26.

To predict the glass transition temperature of (styrene + methyl acrylate + heptyl acrylate) terpolymerization products from Eq. (7.2) one should calculate the fractions of all dyads $\langle P(\bar{M}_i \bar{M}_j) \rangle$ using reactivity ratios (7.4). Combining such calculations with experimental data on glass transition temperatures of homopolymers and alternating binary copolymers

$$T_{11} = 79, \qquad T_{22} = 6; \qquad T_{33} = -68; \qquad T_{12} = 53;$$

$$T_{12} = -27; \qquad T_{23} = -58 \tag{7.5}$$

the theoretical values of T_g were found [310, 201] for all five transparent (i.e. single phase) terpolymers, presented in Fig. 26. These values, as it follows from Table 7.2, are close enough to the experimental results.

All above-mentioned data point out that the developed theoretical approaches can be rather fruitful to predict the transparency and heat resistance of terpolymers.

Table 7.2. Theoretical (T_g^{th}) and experimental (T_g^{ex}) values of glass transition temperatures of terpolymers of various weight compositions prepared at complete conversion $p = 1$ ($x_i^{\prime 0}$ is weight fraction of units $\bar{M}_i$ in terpolymer) [310, 201]

$x_1^{\prime 0}$	$x_2^{\prime 0}$	$x_3^{\prime 0}$	T_g^{th}	T_g^{ex}
0.6	0.1	0.3	30	30
0.6	0.2	0.2	39	34
0.7	0.1	0.2	45	42
0.7	0.2	0.1	53	54
0.8	0.1	0.1	59	59

The success of similar predictions of such characteristics for (styrene + methyl acrylate + methyl methacrylate) terpolymer [310] is one more evidence of the efficiency of the developed approach.

8 Polymerization in Continuous Flow Systems

8.1 Qualitative Peculiarities of Continuous Copolymerization

In industry many chemical processes of polymer synthesis are currently carried out under continuous regimes. Such regimes exhibit a number of important technological advantages compared to the batch regimes. Moreover, the purposeful changes of the statistical characteristics of copolymerization products, as a results of variation of regimes of hydrodynamic stirring in the reactor [10], should be of vital importance to improve some service properties of copolymes.

In order to avoid an undersirable composition inhomogeneity of some copolymers arising from the change of monomer feed composition with the progress of copolymerization, the synthesis of copolymers might be carried out in a continuous stirred tank reactor (CSTR) [324]. In such a process the monomer mixture of composition $\vec{x}^{in}$ is supplied with constant volume rate q to the reactor input, while at its output both composition − monomer $\vec{x}$ and copolymer $\vec{X}$ − remain unchanged under a steady-state regime. Since the monomer concentrations are the same in all points of CSTR and its mean residence time $\theta = V/q$ is large enough compared to the lifetime of the radical, the composition inhomogeneity should have the smallest value which is determined only by the stochastic character of macromolecule growth. It has been convincingly proved [325] by a series of experiments that the sample of copolymer of vinyl pyrrolidone with vinyl acetate obtained in CSTR exhibited a considerably narrower composition distribution than the sample which had the same average composition but was prepared in a traditional batch reactor.

There are several theoretical and experimental investigations [326–336, 17, 15] of copolymerization in CSTR. After the first paper [326] which was rather unsuccessful for practical computations, a series of important works by Litmanovitch et al. [327–329] has been published. In the first of these publications the authors put forth a complete set of kinetic equations in the framework of a terminal model of binary copolymerization performed in CSTR. There has been proposed a rigorous algorithm of solving both the direct and back problems to estimate input and output system characteristics under the steady-state regime. As an example of employing such an algorithm there have been presented procedures for calculating the kinetics, compositions, and molecular weight of copolymerization products for the copolymerization of styrene with methyl methacrylate, using the values of kinetic parameters reported in the literature. In the next paper [328] the authors after comparing the results of theoretical calculations to experimental data on copolymerization of vinyl pyrrolidone with vinyl acetate in CSTR found their good agreement. The further study concerned the various calculation schemes of copolymerization under hydrodynamic regimes different from CSTR [330–333] and processes with participation of three monomers [330, 335]. It is worth noticing the work by Penchev [334] who experimentally proved the reliability of the terminal model to describe the processes of copolymerization of styrene with methyl methacrylate in CSTR. He also determined the values of reactivity ratios for this system from the obtained experimental data. The authors of Ref. [336] put forth a similar approach for the calculation of the above kinetic parameters, but in contrast to Ref. [335] no convincing experimental evidence were provided in its favor. Recently, the principal possibility of the existence of the auto-oscillating regimes of copolymerization in isothermic CSTR has been proved theoretically [17, 15]. Though some problems of continuous copolymerization have been reported in a number of reviews [337, 5, 10, 338, 339, 18, 340], the level of the research in this field (especially of experimental investigations) remains up to now still insufficient despite its obvious practical values.

8.2 Importance of Hydrodynamic Stirring

When such a stirring is absolutely absent in a continuous flow system, as it takes place in the piston reactor (PR), regularities of the batch processes with the same residence time θ are realized. This implies that in order to describe copolymerization in continuous PR one can apply all theoretical equations known for a common batch process having replaced the current time t for θ. As for the equations presented in Sect. 5.1, which do not involve t al all, they remain unchanged, and one can employ them directly to calculate statistical characteristics of the products of continuous copolymerization in PR. It is worth mentioning that instead of the initial monomer feed composition $\vec{x}^0$ for the batch reactor one should now use the vector of monomer feed composition $\vec{x}^{in}$ at the input of PR. In those cases where copolymer is being synthesized in CSTR a number of specific peculiarities inherent to the theoretical description of copolymerization processes arises.

To calculate the statistical characteristics of m-component copolymer prepared in CSTR under the steady-state regime one should determine only the monomer feed composition in the reactor solving a set of algebraic equations [18]:

$$x_i^{in} = (1 - p)\, x_i + p X_i(\vec{x}) \qquad (i = 1, 2, ..., m) \tag{8.1}$$

Then, to calculate copolymer composition, composition distribution, and probabilities of various sequences, one should employ all the equations of statistical theory presented in Sect. 4.

The solution of Eq. (8.1) permits to establish the dependence of the copolymer composition $\vec{X}$ prepared in CSTR on conversion p and $\vec{x}^{in}$. This dependence, generally speaking differs from $\langle \vec{X} \rangle$ calculated for the batch reactor through Eqs. (5.2) and (5.7). This difference would be largest in the range of middle conversions, since in the vicinity of extreme values $p = 0$ and $p = 1$, the copolymer compositions (which are equal to $\vec{\pi}(\vec{x}^{in})$ and $\vec{x}^{in}$, respectively) are the same for both reactors. It is of a certain interest that if $\vec{x}^{in} = \vec{x}^*$, when the monomer feed composition at the reactor input is azeotropic (see Sect. 4.5), the copolymer composition $\vec{X} = \vec{x}^*$

turns out to be independent on conversion as in the case of copolymerization in the batch reactor. It is worth emphasizing that at a fixed value of p, the composition and the microstructure of copolymerization products, prepared under the continuous regime, are independent on the hydrodynamic parameter θ and on the kinetic parameters of initiation as well as termination reactions but depend on the reactivity ratios only.

8.3 Multiplicity of the Steady-State-Regimes

Since Eqs. (8.1) are non-linear, they can have more than one physical solution $\vec{x} = \vec{w}(p)$. It means that at a given monomer feed composition $\vec{x}^{in}$ at the CSTR input and conversion p, one could obtain under a steady-state regime at the reactor output a copolymer which does not have a single unique composition. Which particular one from a certain set of such compositions $\vec{X} = \vec{W}_\nu(p)$ ($\nu = 1, 2, ...$) is realized, depends on the way of approaching the steady-state regime by the system.

The necessary condition of the appearance of such a multiplicity of the steady-state regimes as conversion increases, is when the value $S_1^{(m)}$ becomes zero at some critical value $p = p_c$. This value $S_1^{(m)}$ is a sum of all principle minors of order 1 of the matrix $\mathbf{M}$ of order m, the elements of which:

$$\mu_{ij}(p) = (1 - p)\,\delta_{ij} + p\left.\frac{\partial\pi}{\partial x_j}\right|_{\vec{x} = \vec{w}(p)} \qquad (i, j = 1, 2, ..., m) \qquad (8.2)$$

in the case of the terminal model (2.1) are determined by the derivates of functions $\pi_i(\vec{x})$ (see Sect. 4.1) calculated at the solution $\vec{x} = \vec{w}(p)$ of a set of Eqs. (8.1). When $p \ll 1$, matrix $\mathbf{M}$ is close to the unit matrix $\mathbf{E}$, hence the value $S_1^{(m)} = m$ is positive, independently of the values of vector $\vec{x}^{in}$. Therefore, at all m and monomer feed compositions at the reactor input $\vec{x}^{in}$, a set of Eqs. (8.1) has a single solution when the conversions are rather low. The contribution from the second term of Eq. (8.2) increases with conversion and can result in vanishing of $S_1^{(m)}$ at some $p = p_c$. It is easy to demonstrate that always in binary copolymerization $S_1^{(2)} > 0$, and, consequently one can expect the appearance of the multiplicity of the steady-state regimes in CSTR at a given conversion only in the case of multicomponent systems.

It is worth mentioning that only those stationary solutions of a set of algebraic Eqs. (8.1) can be realized in practice, which are stable. Therefore, we have the problem of determining the stability of the stationary solution by using a relevant set of dynamic equations. For the continuous reactors, in contrast to the batch reactors, such a set is principally more complex since it depends on kinetic parameters of the initiation and termination reactions.

To study the dynamics of the copolymerization it is necessary to know its reduced (i.e., divided by a total monomer concentration) rate. One can consider it to be practically independent on dilution [341, 342] and present it in the form of $\varphi(\vec{x})/t_p$. It is convenient to take the characteristic time of homopolymerization of one of the monomers as a scaling factor. The dependence of the dimensionless function $\varphi(\bar{x})$ on monomer feed composition $\bar{x}$ should be determined from kinetic experiment in the batch reactor since no reliable, well-grounded theory describing the kinetics of chain termination reactions in radical copolymerization is available yet. The dynamics of this process in CSTR is determined by solving the equations of material balance [17, 18]:

$$(1 - p) \frac{dx_i}{d\tau} = (1 - p) \varphi(\bar{x}) (x_i - X_i) + \gamma(x_i^{in} - x_i), \qquad x_i(0) = x_i^0$$

(8.3)

$$\frac{dp}{d\tau} = (1 - p) \varphi(\bar{x}) - \gamma p, \qquad p(0) = 0$$ (8.4)

where $\tau = t/t_p$ is current time t normalized to t_p; $\gamma^{-1} = \theta/t_p$ is the dimensionless Damköhler's parameter. In the limit $\theta \to \infty$ (i.e. $\gamma \to 0$), Eqs. (8.3) and (8.4) are reduce to the well-known set of equations (5.2) describing the drift of monomer feed composition with conversion in the batch reactor. At $t \to \infty$ the trajectories of the dynamic system (8.3)–(8.4) approach the stationary points (SPs), which are determined from the solution of a set of algebraic equations, deduced from Eqs. (8.3)–(8.4) if their left-hand parts equal zero. The complete set of such equations in addition to Eqs. (8.1) involve one more equation:

$$p = \varphi(\bar{x})/[\varphi(\bar{x}) + \gamma]$$ (8.5)

which allows to establish the dependence of stationary conversion in the reactor on its residence time θ. This dependence is determined from the solution of the equation for conversion p, which can be deduced from Eq. (8.5) substituting solution $\bar{x} = \vec{w}(p)$ of a set of equations (8.1) into it. Even if the latter set has a single solution, it is principally possible that the equation for the stationary conversion has several roots in the interval $0 < p < 1$ at a given value of γ. In this case, a multiplicity of the steady-state regimes of copolymerization in CSTR is related to a character of function $\varphi(\bar{x})$, i.e., depends on the initiation and termination reactions. Let us recall that such a dependence is absent in the case of the above-mentioned multiplicity arising when $S_1^{(m)}$ becomes zero.

The stability of an arbitrary SP of the dynamic system (8.3)–(8.4) is studied via the common linearization [204, 205] of a set of equations in the vicinity of a given SP and further examination of the roots of the relevant characteristic equation, similar to (5.11) one for the dynamic system (5.2).

8.4 Binary Copolymerization

In such systems the vector $\bar{x}$ has a single independent component $x_1 = x$, which together with conversion p set up completely the coordinates of the representing point (x, p) of the dynamic system in its phase space, being a square when $m = 2$. Several authors already reported [20, 5, 343, 344] the dependencies of copolymerization rates on the monomer feed compositions for several monomer pairs with known values of r_1 and r_2. This means that the function $\varphi(x)$ is known for each of these pairs, and therefore, the character of the dynamics of the above copolymerization systems depends only on two dimensionless parameters x^{in} and γ.

The dynamic behavior of such systems is determined by the types of stationary points (x_s, p_s) and specific trajectories of a set of equations (8.3) and (8.4). These SPs are found from solution of Eqs. (8.1) (8.5), and their type is established by means of routine procedures of linear analysis [204–206] from the roots of the characteristic polynomial in $\hat{\lambda}$:

$$\hat{\lambda}^2 + \sigma\hat{\lambda} + \Delta = 0 \qquad (8.6)$$

coefficients of which are as follows:

$$\sigma = 2 - p_s + p_s\pi_x^s - \varrho^s, \qquad \Delta = 1 - p_s + p_s\pi_x^s - (1 - p_s)\varrho^s \qquad (8.7)$$

$$\hat{\lambda} = \lambda\frac{(1 - p_s)}{\gamma}, \qquad \pi_x(x) \equiv \frac{d\pi}{dx}, \qquad \varrho(x) \equiv (x - x^{in})\frac{d\ln\varphi}{dx}$$

If coefficients σ and Δ of Eq. (8.6) are positive over the entire range of variation of parameters x^{in} and γ of some system, its dynamics will be trivial since it has only one stable SP. A sufficient condition for this, according to the nonequality $\pi_x(x) > 0$ is the negativity of the value of ϱ^s. This condition might be quite useful to analyze the dynamics of systems (I) and (III) (5.10) with a monotonic dependence of the copolymerization rate on monomer feed composition. In this case the nonequality $\varrho(x) < 0$ is true at all x^{in} and γ, if the fraction of monomer, M_i having a lower rate of homopolymerization in the input is less than the that inside the reactor. This means that, according to the equality $(x - x^{in}) = p(x - X)$, the reactivity ratio r_i of the radical corresponding to monomer M_i obeys the nonequality $r_i < 1$. Therefore, in the course of copolymerization one can observe the continuous enrichment of the monomer mixture with the i-th, less active monomer favoring a decrease of the process rate.

When the residence time θ is large enough, the set of Eqs. (8.3) and (8.4) has a single SP. Its coordinates at $\gamma \rightarrow 0$ approach $(x_s^+, 1)$, where x_s^+ is a single root of equation $\pi(x) = x^{in}$. The stability of this SP, due to the positivity of its parameter Δ (8.7), is determined by a sign of the following expression:

$$\sigma = 1 + \pi_x^+ - \varrho^+ \qquad (8.8)$$

where the superscript of functions π_x and ϱ (8.7) means that they are calculated at $x = x_s^+$. Since the boundaries of the phase space $0 \leq x \leq 1$ and $0 \leq p \leq 1$ are repulsive, this system has at $\sigma < 0$ at least one stable limit cycle. Hence, when $\gamma \rightarrow 0$, the condition $\sigma = 0$ determines with account of expression (8.8) the boundaries of the region of the values x^{in}, where auto-oscillation occurs.

The intervals of the variation of parameters x^{in} and γ, where such a regime exists, have been theoretically established [17] for copolymerization of styrene with

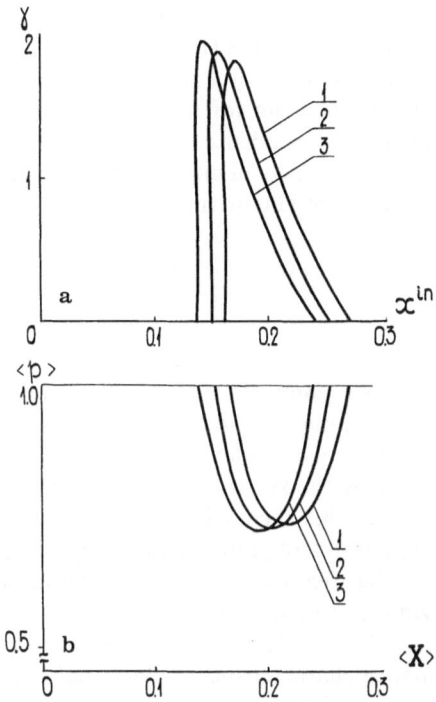

Fig. 27. Boundaries separating the regions of steady-state and auto-oscillation regimes of binary copolymerization in CSTR. Curve 2 is calculated for kinetic parameters resulting in the best approximation of experimental data reported in Ref. [345]; curves 1 and 3 are calculated at the limits of confidence interval of the values of these parameters [17, 6]

butyl acrylate in benzene at 60 °C as an example. The authors of Ref. [345] provided for this system the values of the reactivity ratios, and the dependence of the copolymerization rate on composition. The latter was approximated [17] by a simple function $\varphi(x)$ which described the experimental data well enough. The numerical analysis shows that for this concrete system the value of Δ (8.7) is positive at all values of x^{in} and γ. It testifies to the existence of only one SP. The boundary of the region of these parameters, where SP loses its stability, is determined by the condition $\sigma = 0$ (see Fig. 27). As may be seen from Fig. 27, there is a certain interval of copolymer composition, which could not be realized under a steady-state regime at rather high conversions. It is very essential for the development of new technological regimes of the copolymerization process. At such high conversions the copolymers in this region of compositions could be synthesized in CSTR under the auto-oscillating regime only. The final product, being a result of mixing of copolymer with different instantaneous compositions formed at various moments, should have a considerable composition inhomogeneity. The dispersion $\sigma^2 = \langle \pi^2 \rangle - \langle \pi \rangle^2$ of the composition distribution serves as its quantitative measure. The angular brackets here mean the averaging over the period T of auto-oscillation:

$$\langle p \rangle = \frac{1}{T} \int_0^T p \, dt, \qquad \langle \pi^k \rangle = \frac{1}{\langle p \rangle T} \int_0^T \pi^k p \, dt \qquad (8.9)$$

The values of its dispersion under an auto-oscillating regime of copolymerization of styrene with butyl acrylate in CSTR have been presented [17].

It is worth emphasizing in conclusion of this section that a similar analysis of the dynamic system (8.3), (8.4) could be also carried out for terpolymerization, since by now experimental data are available on the dependence of the copolymerization rate on monomer feed composition for terpolymers of methyl methacrylate and styrene with either diethyl maleate [344], N-vinylpyrrolidone [344], or acrylonitrile [346].

9 Conclusion

As it follows from the present review, a rather complete and experimentally well-grounded quantitative theory of radical copolymerization of an arbitrary number of monomers has been developed. This theory allows one to calculate various statistical copolymers characteristics using the known values of reactivity ratios. The modern stage of the development of this theory is characterized by new approaches applying, for example, the apparatus of graph theory and theory of the dynamic systems which permit to widen the area of theoretical consideration involving the multicomponent copolymerization at high conversions.

The above-mentioned approaches (see Sect. 7) provide us with an opportunity to predict in a number of cases the performance characteristics of the copolymer immediately after their synthesis. The simplicity of these approaches connected with the application of semiempirical relations, makes them very attractive for practical use. However, the price for such a simplicity is a limited area of applicability. In particular, the procedure of predicting the transparency of the multicomponent copolymers, described in Sect. 7, could be effectively used only for those systems where one of the monomers is noticeably different from the others. In the case of arbitrary systems that do not meet the above-mentioned requirement, the boundaries of the transparency regions could be established via a thermodynamical consideration of the polymer-monomer mixture formed at various conversions. The expression for the free energy of such a mixture derived due to this consideration allows to formulate the conditions of the liquid-liquid phase separation, resulting in a transparency loss of the polymerizing system. The development of a rigorous statistical apporach for the description at a molecular level, acccounting for the physical interactions between the components, can be regarded as one of the most important problems of radical copolymerization theory.

Another unsolved fundamental problem of this theory concerns the correct description of copolymerization kinetics which obviously requires a well-grounded expression, from the physicochemical viewpoint, for the rate constant of the bimolecular chain termination reaction. This elementary reaction of interaction of two macroradicals proves to be diffusion-controlled beginning from the very initial conversions, and therefore, its rate in the course of the entire process is controlled by physical, rather than chemical factors. Naturally, the consideration of the kinetics of bulk copolymerization requires different approaches:

a) at low conversions when macroradicals are in diluted solution of their own monomers;

b) at high conversions, when the reaction system is either semi-diluted or a concentrated solution.

Up to now no rigorous quantitative theory describing the kinetics of homopolymerization even of one monomer is available within the framework of modern theory of polymeric solutions. The theory concerning copolymerization will become more complex since the rate constant of a pair of the macroradical recombination could depend not only on their degrees of polymerization but also on the composition and chemical structure of these macroradicals.

In recent years a number of interesting publications appeared, for instance Refs. [347, 296]. Many peculiar regularities of homogenous copopylmerization of traditional monomers were experimentally discovered. These regularities are found to be quite common and could be observed for a number of concrete systems. Moreover, it is essential that these anomalies to which no attention has ever been paid, are pronounced in the region of low conversions, i.e., in the region where the classical copolymerization theory should work best. The reason of such an unusual for this theory phenomenon (dependence of copolymer composition on the degree of copolymerization [347] or invariability of unit distribution in macromolecules with the same compositions, prepared in various solvents at different monomer feed compositions [296]) from the viewpoint of the authors of the above papers is the phenomenon of preferential sorption of monomers in the growing polymeric coils. Therefore, the concentrations of monomers in the coils should have values different from those observed outside the macroradicals in the bulk solution. Obviously, to develop quantitative copolymerization theory accounting for the above-mentioned phenomenon, one should know the monomer feed composition inside each macroradical depending on its molecular weight as well as composition and microstructure. To derive such theoretical dependencies one should consider the conditions of the thermodynamic equilibrium between monomer molecules which are inside and outside the growing polymeric chains.

Hence, the further development of quantitative copolymerization theory, as it ensues from the above-mentioned unsolved problems, is connected with the consideration of physical factors, along with chemical ones, which have a direct influence on process rate and statistical characteristics of the forming products. The future success in such a direction will be connected obviously with the unification of the ideas of chemical copolymerization theory, presented in this review, with the modern concepts of statistical physics both of macromolecules and copolymer solutions.

Acknowledgments: It is a great pleasure for me to express my gratitude to Professor Victor A. Kabanov for his support and continuous attention to my investigations in this field. I am grateful to O. Arzhakova who helped me bring the manuscript into a readable form. I am also indebted to my wife Natasha for her invaluable assistance in preparing the review for publication.

10 References

1. van Krevelen DV (1972) Properties of polymers. Correlations with chemical structure. Elsevier, Amsterdam
2. Korshak VV (1970) Khimicheskoe stroenie i temperaturnye kharacteristiki polimerov. Nauka, Moskva
3. Korshak VV (1977) Raznozvennost polimerov. Nauka, Moskva
4. Gladyshev GP, Popov VA (1974) Radicalnaya polimerizatsia pri glubokikh stepeniakh prevrashenia. Nauka, Moskva
5. Wittmer P (1979) Makromol Chem Suppl 3: 129
6. Kuchanov SI (1978) Metodi kineticheskikh raschetov v khimii polymerov. Khimia, Moskva
7. Seiner JA (1965) J Polym Sci A 3: 2401
8. Friedman E (1965) J Polym Sci B 3: 815
9. Kuchanov SI (1986) Vysokomol soedin B 28: 234
10. Kuchanov SI (1981) Khimich promyshl N 11: 668
11. Wall FT (1944) J Amer Chem Soc 66: 2050
12. Skeist I (1946) J Amer Chem Soc 68: 1781
13. Kuchanov SI, Brun EB (1977) Zh prikl khim 50: 1106
14. Kuchanov SI, Efremov VA, Slinko MG (1986) Vysokomol soedin A 28: 964
15. Kuchanov SI (1987) IUPAC Intern Symp on free radical polymerization. Italy, Genoa p 183
16. Oscilations and Traveling Waves in Chemical Systems (eds) Field RJ, Burger M (1985) Wiley, Interscience New York
17. Kuchanov SI, Efremov VA, Slinko MG (1985) Dokl Acad Nauk SSSR 283: 413
18. Kuchanov SI (1987) Contemporary aspects of the quantitative theory of the radical copolymerization. in: Reactsii v Polimernikh Sistemakh (ed) Ivanchev SS p 107: Khimia, Leningrad
19. Mayo FR, Lewis FM (1944) J Amer Chem Soc 66: 1594
20. Alfrey T, Bohrer JJ, Mark H (1952) Copolymerization. Interscience Publishers, New York
21. Young LJ (1961) J Polym Sci 54: 411
22. Mark H, Immergut B et al (1965) in: Polymer Handbook (eds) Brandrup J, Immergut EH Interscience New York
23. Eastmond GG, Smith EG (1976) in: Free radical polymerization (eds) Bamford CH, Tipper CFH p 333 Elsevier New York
24. Greenley RZ (1980) J Macromol Sci A 14: 427, 445
25. Merz E, Alfrey T, Goldfinger G (1946) J Polym Sci 1: 75
26. Barb WG (1953) J Polym Sci 11: 117
27. Litt M, Bauer FW (1967) J Polym Sci C, N 16: 1554
28. Inaki Y, Hirose S et al (1971) Polymer J 2: 481
29. Kreisel M, Garbatski V, Kohn DH (1964) J Polym Sci A 2: 105
30. Ham G (1954) J Polym Sci 14: 87
31. Natansohn A, Maxim S, Feldman D (1978) Eur Polym J 14: 283
32. Fukuda T, Ma Y-D, Inagaki H (1982) Polymer J 14: 705
33. van der Meer R, Alberti JH et al (1979) J Polym Sci Polym Chem Ed 17: 3349
34. Guillot J, Vialle J, Guyot A. (1971) J Macromol Sci A 5: 735
35. Rounsefell JD, Pittman CU (1979) J Macromol Sci A 13: 153
36. Seiner JA, Litt M (1971) Macromolecules 4: 308
37. Hill DJT, O'Donnell JH, O'Sullivan PW (1982) Progr Polym Sci 8: 215
38. Kabanov VA, Zubov VP, Semchikov YuD (1987) Kompleksno-radicalnaya sopolymerizatsia Khimia, Moskva
39. Lowry GG (1960) J Polym Sci 42: 463
40. Ham GE (1960) J Polym Sci 45: 177
41. Izu M, O'Driscoll KF et al (1972) Macromolecules 5: 90

42. Tsuchida E, Tomono T (1971) Makromol Chem 141: 265
43. Karad P, Schneider C (1978) J Polym Sci, Polym Chem Ed 16: 1137
44. Tudos F, Kelen T, Foldes-Bereznikh T (1975) J Polym Sci, Polym Symp N 50: 109
45. Alfrey T, Goldfinger G (1944) J Chem Phys 12: 322
46. Walling C, Briggs ER (1945) J Amer Chem Soc 67: 1774
47. Ham GE (1964) J Polym Sci A 2: 4191
48. Kutscher RV, Zaitsev JS (1974) Makromol Chem 175: 881
49. Korolev SV, Kuchanov SI (1982) Vysokomol soedin A 24: 638
50. Kang BK, O'Driscoll KF, Howell JA (1972) J Polym Sci A-1 10: 2349
51. Price FP (1962) J Chem Phys 36: 209
52. Stroganov LB (1968) in: Novoe v Metodakh Issledovania Polimerov p 286 Mir, Moskva
53. Price FP (1970) in: Markov chains and monte carlo calculations in polymer science p 187 Marcel Dekker Inc, New York
54. Stokmayer WH (1945) J Chem Phys 13: 199
55. Spencer HG (1975) J Polym Sci, Polym Chem Ed 13: 1253
56. Pittmann CU, Rounsefell TD (1975) Macromolecules 8: 46
57. Cais RE, Farmer RG, Hill DJT, O'Donnell JH (1979) Macromolecules 12: 835
58. Natansohn A, Galea D et al (1981) J Macromol Sci A 15: 393
59. Korolev SV, Kuchanov SI (1982) Vysokomol soedin A 24: 647
60. Kuchanov SI, Korolev SV, Zubov VP, Kabanov VA (1984) Polymer 25: 100
61. Ring W (1963) J Polym Sci B 1: 323
62. Harwood HJ, Ritchey WM (1964) J Polym Sci B 2: 601
63. Ito K, Yamashita Y (1965) J Polym Sci A 3: 2165
64. Coleman BD, Fox TG, Reinmoller M (1966) J Polym Sci B 4: 1029
65. Tosi C (1970) Makromol Chem 138: 299
66. Tada K, Fueno T, Furukawa J (1966) J Polym Sci A-1, 4: 2981
67. Liga A, Negulescu I, Simionescu Cr (1974) Rev Roum Chim 19: 911
68. Georgiev GS (1986) J Macromol Sci A 10: 1063
69. Durgarian AA et al (1986) Vysokomol soedin A 28: 1953
70. Galvan R, Tirrell M (1986) J Polym Sci Polym Chem Ed 24: 803
71. Tosi C, Catinella G (1970) Makromol Chem 137: 211
72. Tosi C (1968) Adv Polym Sci 5: 451
73. Corradini P, Tosi C (1968) Eur Polym J 4: 227
74. Tosi C, Corradini P, Valvassori A, Ciampelly F (1969) J Polym Sci C, N 22: 1085
75. Tosi C (1971) Makromol Chem 150: 199
76. Tosi C (1971) in: NMR basic principles and progress 4: 129
77. Tosi C (1975) Makromol Chem 176: 453
78. Whelan JM (1954) J Polym Sci 14: 409
79. Igarashi S (1963) J Polym Sci B 1: 359
80. Kinsinger JB, Colton D (1965) J Polym Sci B 3: 797
81. Tosi C (1967) Makromol Chem 108: 307
82. Izu M, O'Driscoll KF (1970) J Appl Polym Sci 14: 1515
83. Price FP (1968) J Polym Sci C, N 25: 3
84. Smidsrod O, Whittington SG (1969) Macromolecules 2: 42
85. Mirabella FM (Jr) (1977) Polymer 18: 705
86. O'Driscoll KF (1977) in: Computer in Polymer Science (eds) Mattson JS, Mark HB, MacDonald HC Dekker, New York p 97
87. Motoc I (1977) Math Chem 3: 245
88. Motoc I, Holban S, Cuibotariu D (1977) J Polym Sci, Polym Chem Ed 15: 1465
89. Motoc I, Vancea R, Holban S (1978) J Polym Sci, Polym Chem Ed 16: 1587, 1595
90. Motoc I, Holban S, Vancea R (1978) J Polym Sci, Polym Chem Ed 16: 1601
91. Motoc I, Muscutariu I (1981) J Macromol Sci A 15: 75
92. Motoc I (1981) J Macromol Sci A 15: 85
93. Duever TA, O'Driscoll KF, Reilly PM (1988) J Polym Sci, Polym Chem Ed 26: 965
94. Theil M (1983) J Polym Sci, Polym Chem Ed 21: 633

95. Rakotsi D, Gubka G, Oprescu K (1988) Vysokomol soedin A 30: 123
96. Fuena T, Furukawa J (1964) J Polym Sci A 2: 3681
97. Tosi C, Chioccola G (1973) Makromol Chem 171: 105
98. Tsuchida F, Mishima K et al (1972) J Polym Sci, Polym Chem Ed 10: 3615
99. Harwood HJ, Kodiara Y, Newman DL (1977) in: Computer in Polymer Science (eds) Mattson JS, Mark HB (Jr), MacDonald HC (Jr) Dekker, New York and Basel p 57
100. Mirabella, FM (1977) Polymer 18: 929
101. Motoc I, Muscutariu I, Holban S, Dragomir O (1980) J Polym Sci, Polym Chem Ed 18: 1565
102. Motoc I, O'Driscoll KF (1981) Lecture notes in chemistry 27: 62
103. Kodiara Y, Harwood HJ (1982) in: Computer Application in applied polymer science, (ed) Provder T ACS Sympos Ser 197: 137
104. Tosi C (1970) Eur Polym J 6: 161
105. Ham GE (1983) J Macromol Sci A 19: 699
106. Zilberman EI (1979) Vysokomol soedin B 21: 33
107. Simionescu CrI, Liga A (1976) Rev Roum Chim 21: 1555
108. Seiner JA (1964) J Polym Sci B 2: 985
109. Georgiev GS (1978) J Macromol Sci A 12: 1175
110. Tosi C (1972) Eur Polym J 8: 91
111. Hijmans J (1963) Physica 29: 1
112. Hijmans J (1963) Physica 29: 819
113. Georgiev GS (1979) Makromol Chem 180: 1277
114. Robertson HH (1979) J Inst Math Appl 23: 405
115. Harary F (1969) Graph Theory. Addison-Wesley Publ Comp London
116. Alfrey T, Goldfinger G (1946) J Chem Phys 14: 115
117. Walling C, Seymour D, Wolfstirn KB (1948) J Amer Chem Soc 70: 1544
118. Iwatsuki S, Yamashita Y (1967) Makromol Chem 104: 263
119. Tomescu M, Simionescu Cr (1975) Eur Polym J 11: 453
120. Wittmer P, Demmler K (1975) Makromol Chem Suppl 1: 263
121. Tarasov AI, Tskhay VA, Spasskiy SS (1960) Vysokomol soedin 2: N 11, 1601
122. Braun D, Brendlein W et al (1975) J Macromol Sci A 9: 1457
123. Rios L, Guillot J (1978) J Macromol Sci A 12: 1151
124. O'Driscoll KF (1967) ACS Polym Preprints 8: 205
125. Wittmer P, Hafner F, Gerrens H (1967) Makromol Chem 104: 101
126. Ham GE (1966) J Macromol Chem 1: 93
127. O'Driscoll KF (1968) J Polym Sci C: N 25, 47
128. Slocombe RJ (1957) J Polym Sci 26: 9
129. Kambe Y (1973) J Macromol Sci A 7: 547
130. Cruise DR, Lacombe RG (1970) J Polym Sci A-1 8: 1373
131. Kelen T, Turssanyi B, Disselhoff G (1978) J Macromol Sci A 12: 35
132. Valvassori A, Sartory G (1967) Adv Polym Sci 5: 28
133. Quella F (1989) Makromol Chem 190: 1445
134. Brun EB, Kuchanov SI (1977) Vysokomol soedin A 19: 488
135. Ring W (1967) Makromol Chem 101: 145
136. Zaitsev YuS, Kucher RV, Zubov VP (1977) in: Kinetica i mekhanizm reaktsiy obrazovania polimerov. Naukova, Dumka Kiev p 3
137. Quella F, Braun D (1982) Makromol Chem 183: 2537
138. Ring W (1968) Eur Polym J 4: 413
139. Zaitsev YuS et al (1974) Dokl Acad Nauk Ucr SSR B: N 10, 904
140. Tarasov AI, Tskhay VA, Spasskiy SS (1961) Vysokomol soedin 3: 14
141. Zaitsev YuS et al (1972) Dokl Acad Nauk Ucr SSR B: N 9, 823
142. Ham GE (1983) J Macromol Sci A 19: 693
143. Valvassori A, Sartori G (1962) Rend Ist Lombardo Sci Lettere A 96: 107
144. Crespi G, Valvassori A, Sartori G (1964) in: Copolymerization (ed) Ham G Interscince New York p 295
145. Ham GE (1964) J Polym Sci A 2: 2735

146. Ham GE (1964) J Polym Sci A 2: 4169
147. Ham GE (1964) J Polym Sci A 2: 4181
148. Ham GE (1966) J Macromol Chem 1: 103
149. Ham GE, Lipman DA (1967) J Macromol Sci A 1: 1005
150. Mayo FR (1964) J Polym Sci A 2: 4207
151. O'Driscoll KF (1965) J Polym Sci B 3: 305
152. Tidwell PW, Mortimer GA (1966) J Polym Sci B 4: 527
153. O'Driscoll KF, Tidwell PW, Mortimer G (1967) J Polym Sci B 5: 575
154. Sawada H (1967) J Macromol Sci A 1: 559
155. Ham GE (1967) J Macromol Sci A 1: 561
156. O'Driscoll KF, Izu M (1970) J Macromol Sci A 4: 311
157. Alfrey T (Jr), Bohrer JJ, Mark H (1952) Copolymerization. Interscience Publishers, New York
158. van der Hauw T (1965) J Polym Sci B 3: 715
159. Fordyce RG, Chapin EC, Ham GE (1948) J Amer Chem Soc 70: 2489
160. Jenkins AD (1974) in: Reactivity, mechanism and structure in polymer chemistry, (eds) Jenkins AD and Ledwith AL, Wiley — Interscience Publ, London
161. Bamford CH, Jenkins AD (1961) J Polym Sci 53: 149
162. Bamford CH, Jenkins AD (1963) Trans Faraday Soc 59: 530
163. Jenkins AD (1965) Eur Polym J 1: 177
164. Ham GE (1966) J Macromol Chem 1: 403
165. Pyun CW (1971) J Polym Sci A-2: 9, 577
166. O'Driscoll KF, Ham GE (1967) J Macromol Sci A 1: 1365
167. Tosi C, Campadelli F (1975) Makromol Chem 176: 177
168. Gindin LM et al (1947) Dokl Acad Nauk SSSR 56: 177
169. Spinner IH, Lu BC-Y, Graydon, WF (1955) J Amer Chem Soc 77: 2198
170. Meyer VE, Lowry GG (1965) J Polym Sci A 3: 2843
171. Ring W (1964) Dechema Monograph. 49: 75
172. Kruse RL (1967) J Polym Sci B 5: 437
173. Molau GE (1967) J Polym Sci A-1: 5, 401
174. Miagchenkov VA, Frenkel SYa (1969) Vysokomol soedin A 11: 2348
175. Miagchenkov VA et al (1972) Vysokomol soedin B 14: 693
176. Iziumnikov AL (1976) Vysokomol soedin B 18: 162
177. Iziumnikov AL, Virskiy Yu P (1967) Vysokomol soedin A 9: 1996
178. Miagchenkov VA et al (1968) Dokl Acad Nauk SSSR 181: 147
179. Cantow HJ, Fuchs O (1965) Makromol Chem 83: 244
180. Elias HG (1967) Makromol Chem 104: 142
181. Markert G (1967) Makromol Chem 103: 109
182. Buck M (1970) Angew Makromol Chem 11: 63
183. Mazo LD et al (1971) Vysokomol soedin B 13: 565
184. Rostovtseva SA, Mazo LD, Roskin ES (1976) Izv VUZov, ser Khim i Khim Technologia, 19: 93
185. Riabov SA (1986) Ph D Thesis. NII Khimii i Tekhnologii Polimerov im VA Kargina, Dzerzhinsk
186. Miagchenkov VA, Frenkel SYa (1968) Usp Khimii 37: 2247
187. Litmanovich AD, Iziumnikov AL (1968) in: Novoe v metodakh issledovania polymerov (eds) Rogovin ZA, Zubov VP: s 200
188. Miagchenkov VA, Frenkel SYa (1978) Usp Khimii 47: 1261
189. Miagchenkov VA, Frenkel SYa (1988) Kompozitsionnaya neodnorodnost sopolimerov.: Khimia, Leningrad
190. Stejskal J, Kratochvil P (1978) J Appl Polym Sci 22: 2925
191. Stejskal J, Kratochvil P (1980) J Appl Polym Sci 25: 407
192. Prochazka O, Kratochvil P (1984) J Polym Sci, Polym Lett 22: 501
193. Harwood HJ (1967) ACS Polym Preprints 8: 199
194. Harwood HJ (1968) J Polym Sci C, N 25: 37
195. Chan RKS, Meyer VE (1967) J Macromol Sci A 1: 1089

196. Chan RKS, Meyer VE (1968) J Polym Sci C, N 25: 11
197. Simionescu Cr et al (1967) Makromol Chem 110: 278
198. Balashov AV et al (1972) Izv VUZov, ser Khim i Khim Technologia 15: 1212
199. Simionescu Cr, Liga A (1977) Rev Roum Chim 22: 49
200. Saric K, Janovic Z, Vogl O (1984) J Macromol Sci A 21: 267
201. Kuchanov SI (1989) Makromol Chem, Macromol Symp 26: 371
202. Kuchanov SI (1988) Vysokomol soedin A 30: 2312
203. Kuchanov SI (1989) Matematicheskoe modelirovanie 1: 37
204. Arrowsmith DK, Place CM (1982) Ordinary differential equations. A qualitative approach with applications. Chapmen and Hall, London – New York
205. Spiegel MR (1981) Applied differential equations. Prentice-Hall: 654 p
206. Thompson JMT, Stewart HB (1986) Nonlinear dynamics and chaos. John Wiley and sons
207. Koenig JL (1980) Chemical microstructure of polymer chains. John Wiley and sons
208. Berger M, Kuntz I (1964) J Polym Sci A 2: 1687
209. Ito K, Iwase S, Yamashita Y (1967) Makromol Chem 110: 233
210. Ito K, Yamashita Y (1965) J Polym Sci B 3: 625
211. San Roman J, Madruga EL, Del Puerto MA (1980) Angew. Makromol Chem 86: 1
212. Dhal PK (1986) J Macromol Sci A 23: 181
213. Lin DJ, Petit A, Neel J (1987) Makromol Chem 188: 1163
214. Switala M, Wojtczak Z (1986) Makromol Chem 187: 2411
215. Fineman M, Ross SD (1950) J Polym Sci 5: 259
216. Yezrielev AJ, Brokhina EL, Roskin YS (1979) Vysokomol soedin A 11: 1670
217. Kelen T, Tudos F (1975) J Macromol Sci A 9: 1
218. Joshi RM, Joshi SG (1971) J Macromol Sci A 5: 1329
219. Rodriguez F (1962) J Appl Polym Sci 6: 13
220. Kuo JF, Chen ChY (1982) J Appl Polym Sci 27: 2747
221. Johnston HK, Rudin A (1970) J Paint Technol 42: 429
222. Braun D, Brendein W, Mott G (1973) Eur Polym J 9: 1007
223. Tosi C (1973) Eur Polym J 9: 357
224. Behnken DW (1964) J Polym Sci A 2: 645
225. Tidwell P, Mortimer GA (1965) J Polym Sci A 3: 369
226. Kress AO, Mathias LJ, Cei G (1989) Macromolecules 22: 537
227. Leicht R, Fuhrmann J (1983) J Polym Sci, Polym Chem Ed 21: 2215
228. McFarlane RC et al (1980) J Polym Sci, Polym Chem Ed 18: 251
229. Shtraikhman GA et al (1958) Zh Fiz Khim 32: 512
230. Meyer VE (1966) J Polym Sci A-1: 4, 2819
231. Harwood HJ, Johnston NW, Piotrowski H (1968) J Polym Sci C, N 25: 23
232. Montgomery DR, Fry CE (1968) J Polym Sci C, N 25: 59
233. German AL, Heikens D (1971) J Polym Sci A-1, 9: 2225
234. Makarov KA, Tomashevskaya GM et al (1971) Vysokomol soedin A 13: 239
235. Joshi RM (1973) J Macromol Sci A 7: 1231
236. Tudos F, Kelen T et al (1975) J Macromol Sci A 10: 1513
237. Natansohn A (1978) Brit Polym J 10: 218
238. Yamada B, Itahashi M, Otsu T (1978) J Polym Sci, Polym Chem Ed 16: 1719
239. Van der Meer R, Linssen HW, German AL (1978) J Polym Sci, Polym Chem Ed 16: 2915
240. Kelen T, Tudos F, Turcsanyi B (1980) Polym Bull 2: 74
241. Patino-Leal H et al. (1980) J Polym Sci, Polym Lett Ed 18: 219
242. Watts DG, Linssen HN, Schrijver J (1980) J Polym Sci, Polym Chem Ed 18: 1285
243. Kuo JF, Chen CY (1981) J Appl Polym Sci 26: 1117
244. Cherkezian VO, Litmanovich AD et al (1982) Vysokomol. soedin B 24: 237
245. Hautus FLM, Linssen HN, German AL (1984) J Polym Sci, Polym Chem Ed 22: 3487
246. Hautus FLM, Linssen HN, German AL (1984) J Polym Sci, Polym Chem Ed 22: 3661
247. Chee KK, Ng SC (1986) Macromolecules 19: 2779
248. Plaumann HP, Branston RE (1989) J Polym Sci, Polym Chem Ed 27: 2819
249. Greenley RZ (1975) J Macromol Sci A 9: 505

250. Rounsefell TA, Pittman CV (1977) in: Computers in Polymer Science (eds) Mattson JS, Mark HB, MacDonald HC p 179, Marcel Dekker Inc, New York
251. Laurier GC, O'Driscoll KF, Reilly PM (1985) J Polym Sci, Polym Symp, N 72: 17
252. Batik LI, Basner ME et al (1971) Vysokomol soedin 13: 1133
253. Chen FB, Bufkin BG (1981) J Polym Sci, Polym Chem Ed 19: 3027
254. Moskovskiy SL, Goldin PO, Liubetskiy SG (1982) Vysokomol soedin B 24: 819
255. Hautus FLM, German AL, Linssen HN (1985) J Polym Sci, Polym Lett Ed 23: 311
256. Jaacks V (1972) Makromol Chem 161: 161
257. Burfield DR, Savariar CM (1982) J Polym Sci, Polym Chem Ed 20: 515
258. Barson CA (1984) Eur Polym J 20: 125
259. Tidwell PW, Mortimer GA (1970) J Macromol Sci C 4: 281
260. Joshi RM (1973) J Macromol Sci A 7: 1231
261. Rudin A (1977) in: Computer in polymer science, (eds) Mattson JS, Mark HB, MacDonald HC, p 117, Marcel Dekker Inc, New York – Basel
262. Rudin A (1982) The elements of polymer science and engineering. Academic Press
263. Garcia-Rubio LH et al (1982) in: Computer applications in applied polymer science, (ed) Provder T, p 87, ACS Symposium Ser N 197
264. Rudin A (1984) in: Polymer yearbook, (eds) Elias HG, Pethrick RA, p 193, 1, Harwood Acad Publ
265. O'Driscoll KF, Reilly PM (1987) Makromol Chem, Macromol Symp 10, 11: 355
266. Iwatsuki S, Yamashita Y (1967) Makromol Chem 104: 263
267. Aoyagi J, Shinohara I (1971) J Chem Soc Japan, Industrial Chem Section 74: 784
268. Rudin A, Ableson WR et al (1973) J Macromol Sci A 7: 1203
269. Alexandrov VA, Sautin SN (1977) Zh Prikl Khimii 50: 1368
270. Duever TA, O'Driscoll KF, Reilly PM (1983) J Polym Sci, Polym Chem Ed 21: 2003
271. Guillot J (1971) in: Natural and synthetic high polymers, 4 in Ser. NMR. Basic principles and progress, (eds) Diehl P, Fluck E, Kosfeld R, p 109
272. Harwood H (1965) Angew. Chem 77: 1124
273. Borchardt JK (1985) J Macromol Sci A 22: 1711
274. Chujo R (1966) J Phys Soc Japan 21: 2669
275. Chujo R, Ubara H, Nishioka A (1972) Polymer J 3, 670
276. Makushka RYu, Bayoras GI, Seno M (1988) Vysokomol soedin B 30: 488
277. O'Driscoll KF (1980) J Polym Sci, Polym Chem Ed 18: 2747
278. Rudin A, O'Driscoll KF, Rumack MS (1981) Polymer 22: 740
279. Uebel JJ, Dinan FJ (1983) J Polym Sci, Polym Chem Ed 21: 917 (1983)
280. San Roman J, Madruga EL, Del Puerto MA (1979) Angew Makromol Chem 78: 129
281. Uebel JJ, Dinan FJ (1983) J Polym Sci, Polym Chem Ed 21: 2427
282. Hill DJT, O'Donnell JH, O'Sullivan PW (1982) Macromolecules 15: 960
283. Hill DJT, O'Donnell JH (1987) Makromol Chem, Macromol Symp 10/11: 375
284. Hill DJT, O'Donnell JH, O'Sullivan PW (1987) in: Current topics in polymer science, 1: p 137, Part 3.1, Hanser Publ, Munich, Vienna, New York
285. Hill DJT, Lang AP, O'Donnell JH, O'Sullivan PW (1989) Eur Polym J 25: 911
286. Bevington JC, Huckerby TN, Hutton NW (1982) J Polym Sci, Polym Chem Ed 20: 2655
287. Bevington C, Brener SW, Huckerby TN (1984) Polymer Communications 25: 260
288. Bevington JC, Cywar DA, Huckerby TN, Senogles E, Tirrell DA (1988) Eur Polym J 24, N 7: 699
299. Bevington DA, Cywar DA, Huckerby TN, Senogles E, Tirrell DA (1990) Eur Polym J 26, N 1: 41
290. Jones SA, Prementine GS, Tirrell DA (1985) JACS 107: 5275
291. Cywar DA, Tirrell DA (1986) Macromolecules 19: 2908
292. Prementine GS, Tirrell DA (1987) Macromolecules 20: 3034
293. Prementine GS, Tirrell DA (1989) Macromolecules 22: 52
294. Jones SA, Tirrell DA (1987) J Polym Sci, Polym Chem Ed 25: 3177
295. Prementine GS, Jones SA, Tirrell DA (1989) Macromolecules 22: 770
296. Giese B, Engelbrecht R (1984) Polym Bull 4: 55

297. Tanaka H, Sakai I, Sasai K, Sato T, Ota T (1988) J Polym Sci, Polym Lett Ed 26: 11
298. Tanaka H, Sasai K, Sato T, Ota T (1988) Macromolecules 21: 3536
299. Pittman CU, Rounsefell F (1973) J Polym Sci, Polym Chem Ed 11: 621
300. Pittman CU, Rounsefell TDA (1977) in: Computer in Polymer Science, (eds) Mattson JS, Mark HB, MacDonald HC, p. 145, Marcel Dekker Inc, New York
301. Shawki SM, Hamielec AE (1979) J Appl Polym Sci 23: 3155
302. Harwood HJ (1981) in: Data Processing in Chemistry, (ed) Hippe Z, p 1333, Elsevier Sci Publ, New York
303. Harwood HJ (1987) Makromol Chem, Macromol Symp 10/11: 331
304. McFarlane RC, Reilly PM, O'Driscoll KF (1980) J Polym Sci, Polym Lett Ed 18: 81
305. O'Driscoll KF, Kale LT et al (1984) J Polym Sci, Polym Chem Ed 22: 2777
306. Herz J, Decker-Freyss D, Rempp P (1968) J Polym Sci C, N 16: 4035
307. Johnson M, Karmo TS, Smith RR (1978) Eur Polym J 14: 409
308. Dionisio JM, O'Driscoll KF (1979) J Polym Sci, Polym Lett Ed 17: 701
309. Yamashita Y, Ito K (1969) Appl Polym Symp N 8: 245
310. Kuchanov SI, Orlova ZV et al (1989) Vysokomol soedin A 31: 474
311. Saric K, Yanovic Z, Vogl O (1983) J Polym Sci, Polym Chem Ed 21: 1913
312. Janovic Z, Saric K, Vogl O (1983) J Polym Sci, Polym Chem Ed 211: 2713
313. Janovic Z, Saric K, Vogl O (1985) J Macromol Sci A 22: 85
314. Barton JM (1970) J Polym Sci C, N 30: 573
315. Johnston N (1973) Polym Prepr J Amer Chem Soc 14, N 1: 46
316. Johnston NW (1976) J Macromol Sci-Rev Macromol Chem 14: 215
317. Hirooka M, Kato T (1974) J Polym Sci, Polym Lett Ed 12: 31
318. Daimon H, Okitsu H, Kumanotani J (1975) Polymer J 7: 460
319. Nielsen LE (1953) J Amer Chem Soc 75, N 6: 1435
320. Slavnitskayia NN, Semchikov YuD, Riabov SA (1979) Vysokomol soedin B 21: 23
321. Riabov SA, Slavnitskayia NN et al (1980) Docl Acad Nauk SSSR 253: 118
322. Askill IN, Gilding DK (1981) Polymer 22: 342
323. Lifshits IA, Erenburg EG (1981) Kauchuk i rezina, N 5, 6
324. Wall ET, Delbecq CJ, Florin RE (1952) J Polym Sci 9: 177
325. Agasandian VA, Kudriavtseva LG, Litmanovich AD et al (1967) Vysokomol soedin A 9: 2634
326. Bilous O, Piret EL (1955) AIChEr Journ 1: 480
327. Litmanovich AD, Agasandian VA (1966) Kinetika i kataliz 7, N 2: 309
328. Agasandian VA, Litmanovich AD, Shtern VYa (1967) Kinetika i kataliz 8, N 4: 773
329. Hamidov SS, Musaev UN, Litmanovich AD (1978) Abstracts of short communications of International Symp. on Macromol Chemistry. 2: p 58, Nauka, Moskva
330. Szabo TT, Nauman EB (1969) AIChEr Journ 15: 575
331. O'Driscoll KF, Knorr R (1969) Macromolecules 2: 507
332. Mecklenburgh JC (1970) Can J Chem Eng 48: 279
333. Ray WH, Douglas TL, Godslave E (1971) Macromolecules 4: 166
334. Penchev P (1975) Makromol Chem 176: 1383
335. Glushionok IN (1975) Zh Prikl Khim 48: N 12, 2707
336. Balaraman KS, Kulkarni BD, Mashelkar RA (1982) J Appl Polym Sci 27: 2815
337. Nauman EB (1974) J Macromol Sci-Rev Macromol Chem C 10: 75
338. Hamielec AE, MacGregor JF, Penlidis A (1987) Makromol Chem, Makromol Symp 10/11: 521
339. Biesenberger JA, Sebastian DH (1983) Principles of Polymerization Engineering. Wiley-Interscience Publishers, New York
340. Hamielec AE, Mac Gregor JF, Penlidis A (1989) Copolymerization. p 17 in "Comprehensive Polymer Science" v 3, (eds) Allen G, Bevington JC: Pergamon Press
341. San Roman J, Madruga EL, Del Puerto MA (1983) J Polym Sci, Polym Chem Ed 21: 691
342. Fehervari A, Foldes-Berezshich T, Tudos F (1982) J Macromol Sci Chem A 18: 337
343. Wittmer P (1974) Angew Makromol Chem 39: 35

344. Braun D, Czerwinski WK (1987) Makromol Chem, Makromol Symp 10/11: 415
345. Bradbury JH, Melville HW (1954) Proc Roy Soc, A 222: 456
346. Nikiforov VS, Zaitsev YuS, Bondarenko AV (1980) Ukr Khim Zhurn 46, N 1: 64–68
347. Semchikov Yu D, Smirnova LA et al (1988) Dokl Acad Nauk SSSR 298: 411

Editor: K. Dušek
Received October 30, 1990

Dependence of Viscosity on the Composition of Concentrated Dispersions and the Free Volume Concept of Disperse Systems

L. B. Kandyrin, V. N. Kuleznev*

M. V. Lomonossov Institute of Fine Chemical Technology, 86 Vernadsky Prospect, 117571, Moscow, USSR

By comparing numerous rheological and theoretical equations describing the dependence of the viscosity of dispersions on the concentration with the dependence of the viscosity of liquids on their free volume, the authors concluded that the dependence of viscosity on the concentration can be described in terms of the so-called "free volume of disperse systems", i.e. the difference between the maximum possible content of particles in the system φ_{max} and the actual concentration φ: $(\varphi_f = \varphi_{max} - \varphi)$. This comparison is based on the uniformity of models based on theoretical descriptions of the flow of concentrated dispersions and liquids at molecular level, including the transition region to dispersions losing their fluidity at $\varphi \to \varphi_{max}$ on the one hand, and the glass transition region of liquids, on the other hand. The behaviour of both types of systems in the region close to the limiting region is described by the cluster theories and the percolation theory. Experimental data on the concentration dependence of the viscosity of dispersions are given and it is shown that within a certain concentration range they can be described by equations containing parameter φ_f which are similar to Batschinski's or Doolittle's equations. The approach developed by us has been extended to describe other mechanical properties of highly-concentrated dispersions. Further analogy is drawn between the mobility of particles in concentrated dispersions and the glass transition region. Experimental comparison of the data on the compaction rate of highly-concentrated composites under vibration with the free volume of these composites suggested that transition from the high mobility of particles to the "frozen" state occurs at a constant free volume value of dispersions, 15–17 vol.%. We discuss the advantages of the analog method used for systems similar in structure (or in the method of describing structural changes) in spite of considerable differences in the scale of the phenomena being described.

---

* To whom correspondence should be addressed.

1 Introduction

1.1 The Widespread Use of Concentrated Dispersions

Multiphase, multicomponent compositions are widely used in modern practice. Their properties are defined by the characteristics and content of individual components forming the mixture. To these systems belong metal alloys, polymer mixtures, filled polymer and oligomers, cement solutions and drilling fluids, emulsions, varnishes, adhesives, paints, fibrous and dispersed composites, foodstuffs, biological systems and many others. Quite a large number of them belong to colloidal systems. Their properties depend primarily on the interactions at the interface between the dispersion medium and the disperse phase, which has a large specific surface. To this type of system belong those with a particle size much less than 1 μm. The interphase interaction defines the behavior of the systems. Also widely used are the so-called coarsely-dispersed systems with particle size of $10-10^4$ μm. The properties of systems such as these are to a greater extent determined by the concentration and shape of the disperse phase particles than by their size and nature. The wide occurrence of these systems in nature and engineering requires detailed investigation of their structure and properties. Since the properties and specific features of processing of disperse systems are related to their ability to undergo deformation and to flow, particular attention is given to the investigation of their rheological properties. A large number of theoretical and experimental works are devoted to this problem.

1.2 Types of Disperse Systems

According to the size of the particles forming a disperse phase, dispersions can be classified into finely-dispersed, coarsely-dispersed and systems with a wide particle size distribution (polyfractional). According to the particle concentration, dispersions can be divided into diluted, moderately concentrated and highly concentrated ones. The mechanical behavior of different systems can vary significantly. The shape of disperse phase particles makes a large contribution to the behavior of disperse systems. There are disperse systems with particles of regular shape (spheres, ellipsoids, cylinders), of irregular shape (granular materials obtained by grinding), of elongated shape (short and long fibers). There also exist hybrid disperse systems, including different types of particles. It should be noted that the state of aggregation of disperse phase and dispersion medium and their viscosity ratio play an important role in the behavior of disperse systems since their general classification also includes dispersions with solid particles, emulsions with liquid drops and foams with gas inclusions.

1.3 Subjects Under Consideration

These are disperse systems with relatively large solid particles with a size of between 1 and 10^3 μm. Of principal interest are concentrated and highly-concentrated dispersions with particles of irregular shape distributed in a liquid matrix. Such systems, as a rule, show non-Newtonian properties in flow and are characterized by a yield point. Also, partial settling of particles may take place in these systems. No sufficiently detailed theoretical models have so far been developed to describe their mechanical behavior. This is due to the complexity of their structure and the necessity of taking many key factors into account simultaneously. For description of the properties of the systems under consideration we have attempted to use a new approach that does not claim to describe their deformation behavior completely, but is, in our opinion, of conceptual interest.

1.4 The Structure of the Survey

The introduction forms Chapter 1 of this survey. Chapter 2 deals briefly with various approaches to the description of the concentration dependence of the viscosity of disperse systems, including the transition region from fluid to solid-like systems. Chapter 3 describes viscosity from the standpoint of the free volume theory and the specific features of the transition from mobile to glasslike systems. Chapter 4 presents the concept of the free volume of disperse systems developed by us as well as the results of experiments illustrating it. Chapter 5 contains the pertinent generalizations and conclusions.

2 The Dependence of the Viscosity of Disperse Systems on Concentration and Their Flow Theories

2.1 Phenomenological Approaches and Empirical Formulas

Investigation of the dependence of the viscosity of dispersions on concentration has long attracted the attention of scientists. The hydrodynamic model developed by Einstein [1] has come to be considered classical in this direction of research. This model has led to the expression:

$$\eta_{rel} = 1 + 2.5\varphi \tag{1}$$

where η_{rel} is the relative viscosity of dispersion and φ is the volume fraction of disperse phase particles. The extreme simplicity of the final formula results from significant simplification, so that the applicability of Eq. (1) became is to strongly diluted dispersions. Due to practical requirements, Eq. (1) was refined and the range of its applicability was extended. These attempts led to an improvement of the hydrodynamic model proposed by Einstein. The theoretical approaches developed on the basis of the hydrodynamic model will be considered in the next section. Quite a few attempts were also made to obtain phenomenological expressions and empirical formulas based on Eq. (1). To these belongs Hatschek's equation [2], in which, to extend the range of applicability of Eq. (1), the value of Einstein's coefficient was increased:

$$\eta_{rel} = 1 + 4.5\varphi. \tag{2}$$

Another approach to enhance the dependence of viscosity on concentration is the application of the equations derived by Baker [3]:

$$\eta_{rel} = (1 + K_E\varphi)^n \tag{3}$$

and Hess [4]:

$$\eta_{rel} = \left(1 + \frac{K_E}{n}\varphi\right)^n, \tag{4}$$

where K_E is Einstein's coefficient ($K_E = 2.5$), and n is a constant; or by Orr and Blocker [5]:

$$\eta_{rel} = 1 + a\varphi^K, \tag{5}$$

where a and k are empirical constants.

The range of applicability of the above equations also proved to be quite limited. One more approach to the description of the dependence of viscosity on concentration was developed, and it was based on the Arrhenius equation proposed as early as the nineteenth century [6]:

$$\eta_{rel} = \exp{(KC)}, \tag{6}$$

where K is an empirical constant and C is the weight of solute per cm^3 of solvent. This equation also proved sufficiently suitable for the description of the dependence of viscosity on the concentration of disperse systems. Of the same form is the expression proposed by Weltmann and Green [7]:

$$V = (\eta_0 + A)\exp{(B\varphi)}, \tag{7}$$

where V is the plastic viscosity of the system, η_0 is the viscosity of the dispersion medium, φ is the volume fraction of solid matter, A and B are empirical constants. An attempt to define them led to Eq. [8]:

$$\eta_{rel} = \exp{[C \ln{(\eta_1/\eta_0)}]}, \tag{8}$$

where η_1 and η_0 are the viscosity of disperse phase and dispersion medium, respectively, and c is the dispersion concentration. For a description of the concentration dependence of viscosity of suspensions of fibers in polymer solutions Nicodemo and Nicolais [9] recommended the use of the empirical equation similar to Eq. (6):

$$\eta_{rel} = \exp (8.52\varphi).$$

(9)

Recently, Tangsathitkulchai and Austin have deduced a much more elaborate expression for viscosity of concentrated dispersions with a natural particle size distribution [10]:

$$\lg \eta_{rel} = \left\{4.26 + \frac{1}{1 + \left[\dfrac{0.06}{(m - 0.7)^2}\right]^{4.7}}\right\} \left(\frac{\varphi}{1 - \varphi}\right)^{1.1 + \exp(-14.2m^{2.6})},$$

(10)

where m is the modulus of the size distribution of dispersion particles and φ is their volume fraction.

Quite a number of empirical equations describe the concentration dependence of the relative fluidity of dispersions $(1/\eta_{rel})$. Such equations are also to be found in a very early work by Hatschek [2], who related the fluidity to the cube root of the volume fraction of the disperse phase:

$$1/\eta_{rel} = 1 - \varphi^{1/3}$$

(11)

and in Ting and Luebbers's work [11], who on the basis of the simplest equation

$$1/\eta_{rel} = 1 - \varphi$$

(12)

obtained the empirical equation dependent on the viscosity and density of dispersion medium:

$$1/\eta_{rel} = 1 - \frac{\varphi}{0.460 - 1.58 \cdot 10^{-3} (\eta_0/R)^{0.469}},$$

(12a)

where η_0 is the viscosity of dispersion medium, R is the relative density of dispersion medium and disperse phase.

All the above formulas are one-parameter equations, i.e. they relate the dispersion viscosity only to the volume fraction of particles contained in it. This limits the range of applicability of the equations to not very high dispersion concentrations. To take account of the influence of the structure of concentrated dispersions on their rheological behavior, Robinson [12] suggested that the viscosity of dispersions is not only propertional to the volume fraction of solid phase, but is also inversely proportional to the fraction of voids in it. (At about the same time Mooney [40], who proceeded from a hydrodynamic model, arrived, using theoretical methods, at the same conclusion). Robinson's equation contains the "relative sedimentation volume" value $-S'$, which depends on the particle size distribution of the dispersion and is constant for the given filler,

$$\eta_{rel} = 1 + \frac{K\varphi}{1 - S'\varphi}, \tag{13}$$

where K is a constant. By analysis of Robinson's equation, Mori and Ototake concluded that the relative sedimentation volume is inversely proportional to the maximum possible filler concentration corresponding to a closely-packed state ($S' \equiv 1/\varphi_{max}$). Mori and Ototake [13] proposed a refined form of Robinson's equation:

$$\eta_{rel} = 1 + \frac{\bar{d}S}{2(1/\varphi - 1/\varphi_{max})}, \tag{14}$$

where $\bar{d}$ is the effective mean diameter of disperse phase particles and S is their volume specific surface. Attempts to relate the relative viscosity of dispersions to the void content of fillers were also made by Trawinski [14]:

$$\eta_{rel} \sim \left(\frac{\varepsilon}{\varepsilon_{cr}}\right)^n, \tag{15}$$

where ε is the filler porosity, ε_{cr} is the filler porosity at a certain (critical) filler content, and n is a constant depending on the particle size, and also by Leva [15] in a more complex form:

$$\eta_{rel} \sim \left[\frac{\varepsilon^3}{(1 - \varepsilon)^2}\right]^n. \tag{16}$$

An impressive review of different formulas for the dependence of viscosity on concentration was published by Rutgers [16]. Orr and Dalla Valle [17] derived a formula relating the dispersion viscosity to the filler porosity in a different form:

$$\eta_{rel} = \left(\frac{\varepsilon - \varepsilon_f}{1 - \varepsilon_f}\right)^{-1.8}, \tag{17}$$

where ε_f is the filler porosity corresponding to its closest packing.

Equations taking into account the specificity of packing of filler particles were proposed in a somewhat different form by Hawksley [18]

$$\eta_{rel} = \frac{(1 - \varphi)^2}{V/V_0} \tag{18}$$

and by Kynch [19]

$$\eta_{rel} = \frac{(1 - \varphi)}{V/V_0}.$$ (19)

The defining parameter in these equations is the relation of the sedimentation rates of an individual particle (V_0) to that of a particle in dispersion (V). There are indications (see [29]) that formulas of type (18) are applicable to the description of the rheological behavior of dispersions up to high concentrations.

It proved later that it is very convenient to relate the volume fraction of the disperse phase in dispersions to its maximum packing fraction (φ/φ_{max}). This is evidenced by the large number of equations containing this parameter. Among these a different form of Eq. (17) can be mentioned:

$$\eta_{rel} = (1 - \varphi/\varphi_{max})^{-1.8}$$ (17a)

and also Landell's equation [20]:

$$\eta_{rel} = (1 - \varphi/\varphi_{max})^{-2.5}$$ (20)

and that of Pliskin-Tokita [21]:

$$\eta_{rel} = (1 - \varphi/\varphi_{max})^{-N},$$ (21)

where the constant N depends on the degree of orientation of disperse phase particles.

Of a slightly different form is the empirical equation of Eilers [22]:

$$\eta_{rel} = \frac{25}{16}\left[\frac{\varphi}{1 - (\varphi/\varphi_{max})^2}\right]$$ (22)

and Fedors' equation [23]:

$$\eta_{rel} = \left(1 + \frac{1.25\varphi}{\varphi_{max} - \varphi}\right)^2.$$ (23)

The equation proposed by Chong, Cristiansen and Baer [24] for the description of the rheological behavior of dispersions with a bimodal particle size distribution also includes the normalized value of the volume fraction of the disperse phase:

$$\eta_{rel} = \left[1 + 0.75\frac{\varphi/\varphi_{max}}{1 - \varphi/\varphi_{max}}\right]^2.$$ (24)

Pal and Rhodes [25] report that experimental results are described well using the relation (φ/φ_{100}) in which φ_{100} is the volume fraction of particles corresponding to a 100-fold increase of the relative viscosity of emulsions.

$$\eta_{rel} = \left[1 + \frac{\varphi/\varphi_{100}}{1.187 - \varphi/\varphi_{100}}\right]^{2.492}. \tag{25}$$

An equation relating the viscosity of dispersions to the relation φ/φ_{max} in an exponential form was proposed by Johnston and Brower [26]:

$$lg\,\eta_{rel} = [1.33 - 0.84(\varphi/\varphi_{max})]\,[\varphi/(\varphi_{max} - \varphi)]. \tag{26}$$

Most of the works mentioned above contain substantial experimental material, many times cited and used to substantiate various theoretical models (to be reviewed in the next section).

2.2 Theoretical Models

The majority of theories describing the concentration dependence of viscosity of diluted and moderately concentrated disperse systems is based on the hydrodynamic approach developed by Einstein [1]. Those theories were fairly thoroughly analyzed in the reviews written by Frish and Simha [28] and by Happel and Brenner [29]. In a fairly large number of works describing the dependence of viscosity on concentration the final formulas are given in the form of a power series of the volume concentration of disperse phase particles $-\varphi$. An equation of this type was also proposed by Kunitz [30]:

$$\eta_{rel} = 1 + K\varphi + \frac{K^2}{2!}\varphi^2 + \frac{K^3}{3!}\varphi^3 + \dots. \tag{27}$$

The experimental dependence was described quite adequately by Eq. (27) with the value of the empirical constant $K = 4.15$. A similar dependence was proposed by Brinkman [31]:

$$\eta_{rel} = 1 + [\eta] + 0.7[\eta]^2\,G_w + 0.42[\eta]^3\,G_w^2 + \dots, \tag{28}$$

where $[\eta]$ is the characteristic viscosity of spherical particles depending on their density, G_w is the weight fraction of particles in dispersion. Brinkman's equation is similar to that of Heller [32], who transformed Einstein's equation (1) to the form corresponding to Huggins' equation:

$$\eta_{rel} = 1 + [\eta]\,\varphi(1 + K_1[\eta]\,\varphi + K_2[\eta]\,\varphi^2 + \dots), \tag{29}$$

where $[\eta]$ is the limit of relation η_{sp}/φ at $\varphi \to 0$, and K_1 and K_2 are Huggins constants which do not depend on $[\eta]$.

One of the first attempts to take into account the first order hydrodynamic interactions in diluted suspensions theoretically was the exponential equation derived by Guth and Simha [33]:

$$\eta_{rel} = 1 + 2.5\varphi + 14.1\varphi^2. \tag{30}$$

Later, taking into account the volume occupied by dispersion particles of not very diluted dispersions led to a more accurate value of the coefficient of the quadratic term:

$$\eta_{rel} = 1 + 2.5\varphi + 12.6\varphi^2. \tag{30a}$$

By analysis of the change in hydrodynamic interactions due to the presence of particle doublets and triplets, in dispersions, Vand [34] obtained the expression:

$$\eta_{rel} = 1 + 2.5\varphi + 7.349\varphi^2 + 0(\varphi^3), \tag{31}$$

where 0 is Oseen's operator, or in the numerical form:

$$\eta_{rel} = 1 + 2.5\varphi + 7.17\varphi^2 + 16.2\varphi^3. \tag{31a}$$

Kynch's [35] calculations led to similar expressions:

$$\eta_{rel} = 1 + 2.5\varphi + 7.5\varphi^2 \tag{32}$$

and so did those of Lewis and Nielsen [36], in which account was also taken of the presence of particle aggregates in flowing dispersions:

$$\eta_{rel} = 1 + 2.5\varphi + 7.031\varphi^2 + 37.371\varphi^3. \tag{33}$$

Petersen and Fixman [37] obtained a similar expression, which was later refined by Bedeaux et al. [38] and presented in the form:

$$\eta_{rel} = 1 + 2.5\varphi + 4.8\varphi^2. \tag{34}$$

A most complete theoretical calculation of hydrodynamic interactions between dispersion particles led Batchelor and Green [39] to the expression:

$$\eta_{rel} = 1 + 2.5\varphi + (5.2 \pm 0.3)\,\varphi^2 + 0(\varphi^3), \tag{35}$$

or in the numerical form:

$$\eta_{rel} = 1 + 2.5\varphi + (7.6 \pm 0.8)\,\varphi^2. \tag{35a}$$

According to Happel and Brenner [29], the value of Einstein's coefficient ($K_E = 2.5$) has not been completely proved and therefore it is difficult to obtain theoretically the exact value of the coefficient before the quadratic term.

Along with the polynomial forms of the dependence of viscosity of dispersions on concentration, the theoretical expression in the exponential form also came to be used. Such as, for example, Vand's formula [34] obtained for aggregated dispersions:

$$\eta_{rel} = \exp\left[\frac{K_1\varphi + \eta_2(K_2 - K_1)\,\varphi^2}{1 - Q\varphi}\right], \tag{36}$$

where K_1 is Einstein's coefficient ($K_1 = 2.5$), K_2 is the form factor for particle aggregates depending on the form of packing of particles in dispersion, η_2 is the constant of the existence of aggregates in time ($\eta_2 \simeq 4$), Q is the constant of hydrodynamic interaction ($Q = 0.609$).

The most widely used equation describing the dependence of viscosity on concentration is Mooney's formula [40]:

$$\eta_{rel} = \exp\left(\frac{2.5\varphi}{1 - K'\varphi}\right), \tag{37}$$

where the term K' takes account of the so-called "self-constraint" of disperse phase particles and is equal to $1/\varphi_{max}$. As pointed out by Rutgers [16], Mooney's equation (37) describes the dependence of the viscosity of dispersions on particle concentration very well most of the time. It is valid, however, only for the concentration range far removed from the maximum filler content of dispersions φ_{max}. On the basis of Mooney's equation, Brodnyan [11] derived a formula describing the concentration dependence of the viscosity of elliptical dispersion particles:

$$\eta_{rel} = \exp\left[\frac{2.5\varphi + 0.399(p - 1)^{1.48}}{1 - K'\varphi}\,\varphi\right], \tag{38}$$

where p is the relation between the large and small ellipsoid axes. Brodnyan [41] points out that Eq. (38) describes the experimental results better than Simha's [42] theoretical equation does:

$$\eta_{rel} = 1 + \frac{14}{15}\varphi + \frac{p^2}{5}\left[\frac{1}{3(\ln 2p - K)} + \frac{1}{\ln 2p - K + 1}\right]\varphi, \tag{39}$$

(where p is the relationship between ellipsoid axes and K is a constant valid at $p > 5$) and also better than the equations obtained by Kuhn and Kuhn [43]:

$$\eta_{rel} = 1 + 2.5\varphi + \frac{32}{15\pi}\left(\frac{1}{p} - 1\right)\varphi - 0.628\,\frac{(1/p - 1)}{(1/p) - 0.75}\,\varphi$$
$$\text{at}\quad 0 < p < 1 \tag{40}$$

$$\eta_{rel} = 1 + 2.5\varphi + 0.4075(p - 1)^{1.508}\,\varphi \quad \text{at}\quad 1 < p < 15 \tag{40a}$$

$$\eta_{rel} = 1 + 1.6\varphi + \frac{p^2}{5}\left[\frac{1}{3(\ln 2p - 1.5)} + \frac{1}{\ln 2p - 0.5}\right]\varphi$$
$$\text{at}\quad p > 15. \tag{40b}$$

An equation similar in form to Mooney's equation was derived by Kunnen [44], who proceeded from the additivity of the reciprocal values of the activation energy of the viscous flow for binary solutions, emulsions and dispersions:

$$\eta_{rel} = \exp\left(\frac{Kf\varphi}{1 - f\varphi}\right), \tag{41}$$

where $K = \ln(\eta_1/\eta_e)$, η_e is the viscosity of the system at the intersection point of the dependences $\ln\eta - 1/T^0K$ for the mixture components,

$f = 1 - \dfrac{K}{\ln(\eta_2/\eta_1) + K}$, η_2 and η_1 are the viscosities of disperse phase and dispersion medium of the two solutions being mixed. Kunnen points out that the value of $f > 1$ is characteristic for dispersions of nonspherical particles, $f = 1$ for solid spherical particles and $f < 1$ for liquid mixtures and solutions. Thomas [45] proposed that an additional term should be introduced into the conventional equation describing the dependence of viscosity on dispersion concentration in the form of a polynomial with different powers of φ, the term which according to Eyring's theory [85] corresponds to the transfer of particles from one position to another. This resulted in the equation:

$$\eta_{rel} = 1 + 2.5\varphi + 10.05\varphi^2 + A\exp(B\varphi), \tag{42}$$

where A and B are constants, or:

$$\eta_{rel} = 1 + 2.5\varphi + 10.05\varphi^2 + C_1\exp\left(\frac{C_2\varphi}{1 - C_3\varphi}\right), \tag{42a}$$

where C_1, C_2, C_3 are constants. This refinement of the equation describing the dependence of the viscosity on concentration enabled Thomas to describe with the use of Eq. (42) the experimental dependences up to $\varphi = 0.6$.

The theoretical equations relating relative fluidity $(1/\eta_{rel})$ to dispersion concentration also became widely known. Among them is Brinkman's equation [31], valid for dispersions of particles with a wide particle size distribution:

$$1/\eta_{rel} = (1 - \varphi)^{2.5}. \tag{43}$$

A modification of this equation for dispersions with a monodisperse particle distribution was proposed by Roscoe [46]:

$$1/\eta_{rel} = (1 - 1.35\varphi)^{2.5}. \tag{44}$$

The coefficient before the volume fraction of disperse phase particles in Eq. (44) was found on the basis of Vand's theory [34] of the immobilization of the dispersion medium in particle assembles; it is equal to $\dfrac{3\sqrt{2}}{\pi}$. Gillespie [47] suggested a different value of the exponent in Brinkman's equation:

$$1/\eta_{rel} = (1 - \varphi)^{(K_E + K_p\varphi)},\qquad(45)$$

where K_E is Einstein's coefficient ($K_E = 2.5$), K_p is a constant. In the equation proposed by De Bruijn [48] the concentration dependence of dispersion viscosity is of polynomial form:

$$1/\eta_{rel} = 1 - 2.5\varphi + 1.552\varphi^2.\qquad(46)$$

An equation similar to Eq. (46) was obtained by Bedeaux [49]:

$$1/\eta_{rel} = 1 - 2.5\varphi.\qquad(47)$$

It should be noted that an expression similar to (47) was recommended by Budianski [50] and Hill [51] for description of the dependence of the relative modulus of disperse composites on concentration. A more elaborate expression was obtained by Lee [53], who used Vand's concepts of the immobilized liquid for the description of the behavior of dispersions in terms of the mechanics of three bodies:

$$1/\eta_{rel} = (1 - \varphi)^{(2.5 + 1.92\varphi + 7.739\varphi^2)}.\qquad(48)$$

Equations, similar to Hatschek's empirical equation (11) can be found in calculations by Ishai and Cohen [54] or Narkis [55] for describing the dependence of relative modulus on concentration for disperse composites:

$$1/\mu_{rel} = 1 - \varphi^{1/3},\qquad(49)$$

where μ_{rel} is the relative modulus.

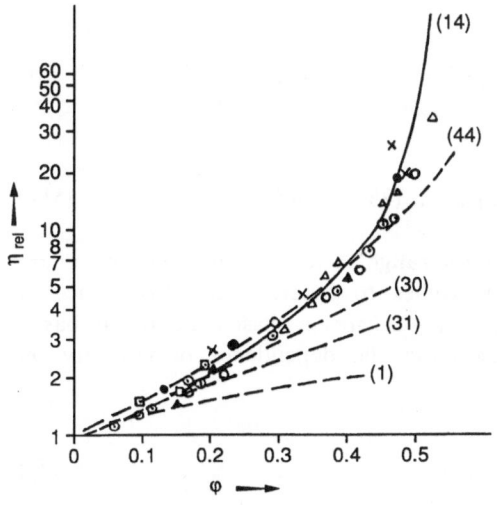

Fig. 1. Concentration dependences of the viscosity of dispersions according to Mori-Ototake [13]. Numbers on the curves correspond to numbers of equations in the text. Experimental values are given from data in [34], [12], [5], [17]

Figure 1 shows a comparison, published by Mori and Ototake [13], of the experimental dependences of viscosity on concentration of dispersions of solid particles based on the data of Vand [34], Robinson [12], Orr and Blocker [5], Dalla Valle and Orr [17] with the theoretical equations based on the hydrodynamic approach used by Einstein (1), Simha (30), Vand (31), Roscoe (44) and the phenomenological equation of Mori and Ototake (14). A more complicated form of the theoretical dependence, naturally makes it possible to describe experimental results over a wider range, but for concentrated dispersions most of theoretical equations remain inapplicable.

Analysis of the hydrodynamic interactions of many particles in a laminar flow, carried out by Saito [56] showed that in view of the complexity of the physical picture of interactions in many body systems introduction of Stokes approximations in a theoretical consideration of the flow of dispersions can lead to incorrect results. For the laminar flow Saito proposed a formula containing a power series:

$$\eta_{rel} = 1 + 2.5\varphi + K(2.5\varphi)^2 . \tag{50}$$

Nagatani [57] arrived at a similar equation: he presented a generalized theory describing slow viscous flow of suspensions in terms of statistical continuum mechanics.

Attempts to describe the unlimited increase of the viscosity of dispersions and emulsions observed when their concentrations approach the maximum values (φ_{max}) meet great theoretical difficulties. Various approaches were developed to overcome these difficulties. Thus, for example, Russel et al. [58] suggested that account should be taken of the Brownian motion of particles in colloidal dispersions in the form of a hydrodynamic contribution. They showed that this contribution which is to be taken into account in considering a slow flow (with slow shear rates $\dot{\gamma}$), increases considerably with increasing dispersion concentration. For a description of the dependence of viscosity on concentration the above authors obtained an exact equation only in the integral form. At low shear rates it gives the following power series:

$$\eta_{rel} = 1 + 2.5\varphi + (4 \pm 2)\,\varphi^2 + (42 \pm 10)\,\varphi^3 \tag{51}$$

and at high shear rates:

$$\eta_{rel} = 1 + 2.5\varphi + (4 \pm 2)\,\varphi^2 + (25 + 7)\,\varphi^3 . \tag{51a}$$

These two expressions are valid only in the range of not very high disperse phase concentrations ($\varphi \leqq 0.35$). Some approaches to describe the dependence of viscosity on concentration in the range of high disperse phase contents are based on Taylor's well-known formula describing the dependence of viscosity on concentration for emulsions [59]:

$$\eta_{rel} = 1 + \frac{2\eta_1 + 5\eta_2}{2(\eta_1 + \eta_2)}\,\varphi , \tag{52}$$

where η_1 and η_2 are the viscosities of the dispersion medium and the disperse phase of emulsions. Thus, for example, the full expression, obtained on the basis of Eq. (52) by Nagatani [57] in terms of statistical continuum mechanics is in a fairly complex form:

$$\eta_{rel} = 1 - \frac{5(\eta_1 - \eta_2)}{(3\eta_1 + 2\eta_2)}\varphi + \frac{25(\eta_1 - \eta_2)^2}{(3\eta_1 + 2\eta_2)^2}\varphi^2$$
$$- \frac{125(\eta_1 - \eta_2)^3}{(3\eta_1 + 2\eta_2)^3}\varphi^3 + \dots . \tag{53}$$

Similar dependences for describing the viscosity of emulsions on the basis of the volume averaging method were derived by Mellema and Willemse [60]. They took into account the contribution to the effective viscosity of the interphase tension and obtained a general expression, different from (52):

$$\eta_{rel} = 1 - \frac{5(\eta_1 - \eta_2)}{3\eta_1 + 2\eta_2}\varphi . \tag{54}$$

This expression coincides with the first term of equation (53) and can be expanded into the following series: for suspensions (at $\eta_2 \to \infty$)

$$\eta_{rel} = 1 + 5/2\varphi + 5/2\varphi^2 + \dots \tag{54a}$$

for foams (at $\eta_2 \to 0$)

$$\eta_{rel} = 1 - 5/2\varphi + 10/9\varphi^2 + \dots . \tag{54b}$$

Bedeaux [49], who proceeded from a analogy between the viscous and dielectric properties of dispersions obtained a slightly different result:

$$\eta_{rel} = 1 + 5/2\varphi \frac{\eta_2 - \eta_1}{\eta_2 + 3/2\eta_1}\left[1 - \varphi\left(\frac{\eta_2 - \eta_1}{\eta_2 + 3/2\eta_1}\right)\right]^{-1} . \tag{55}$$

At $\eta_2 \to \infty$ this equation changes to an expression, similar to a different form of Saito's formula [56]:

$$\eta_{rel} = 1 + 5/2\frac{\varphi}{1 - \varphi} . \tag{56}$$

An expression of the same form was obtained by Felderhof [61] when he analysed the flow perturbation caused by a single particle in terms of the force multipoles theory. Bedeaux [49] pointed out that when the disperse phase concentration approaches the maximum value (φ_{max}), the correlation interaction between particles increases. Then the following expression is valid:

$$\frac{\eta_2 - \eta_{eff}}{\eta_2 + 3/2\eta_{eff}} \varphi + \frac{\eta_1 - \eta_{eff}}{\eta_1 + 3/2\eta_{eff}} (1 - \varphi) = 0, \tag{57}$$

where η_{eff} is the effective viscosity of dispersion. At $\eta_2 \to \infty$, i.e. for dispersions containing solid spheres this expression assumes the form (47). Bedeaux also believed that greater similarity of the theoretical and experimental viscosity values for highly concentrated emulsions can be reached by assuming for calculations that the value of the relation η_2/η_1 can attain fairly high, but not infinite, values. The best agreement between the theoretical and experimental values was obtained at $\eta_2/\eta_1 \simeq 200$.

Further analysis of the viscosity behavior of concentrated dispersions of solid spheres with regard to the effective self-consistence at all orders of particle interaction led Bedeaux [62] to a more complex expression:

$$\frac{\eta_{rel} - 1}{\eta_{rel} + 3/2} = \varphi[1 + S(\varphi)], \tag{58}$$

where $S(\varphi)$ accounts for the interaction and correlation between two and more spheres in dispersion. Bedeaux assumes an exponential form of the dependence $S(\varphi)$ for which the second order is the most likely one. At low shear rates (in the case of laminar flow) comparison of theory with experiment gave the following expression:

$$S(\varphi) = (3.08 \pm 0.11)\,\varphi - (3.15 \pm 0.21)\,\varphi^2. \tag{59}$$

Felderhof et al. [63] obtained an expression similar to (58) by refining Saito's equation (56) in terms of the cluster theory.

Theoretical studies of the flow of concentrated dispersions in terms of statistical theories are also reported in the works of Lundgren [64], Beenakker [65], and others. On the other hand, using expressions obtained in describing the viscosity of concentrated emulsions (for example, equation (54)), on the basis of the effective medium theory, Bedeaux derived the following equation:

$$\frac{\eta_{eff} - \eta_1}{\eta_{eff} + 3/2\eta_1} = \varphi \frac{\eta_2 - \eta_1}{\eta_2 + 3/2\eta_1}. \tag{60}$$

At $\eta_2 \to \infty$ (i.e. for dispersions with solid particles) it changes to an expression similar to Saito's formula:

$$\eta_{rel} = \frac{1 + 3/2\varphi}{1 - \varphi}. \tag{60a}$$

A relation similar to (60) was obtained by Ladd [66] when he investigated the creep flow of highly concentrated dispersions taking into account by a numerical method the hydrodynamic interaction between particles:

$$\eta_{rel} = \frac{1 + 5/2\varphi\gamma}{1 + 10\varphi\gamma\chi},$$ (61)

where χ is the geometric factor, which is 1/10 for spherical particles, γ is the dipole-dipole friction tensor component. For $\varphi_{max} = 0.45$, for example, in a numerical calculation for an assembly of 108 particles, (61) changes to the following equation:

$$\eta_{rel} = 1 + \frac{5\varphi}{1 - 2\varphi + 0.88\varphi^2}.$$ (61a)

Expression (61) is claimed to hold over the whole range of dispersion concentrations from diluted systems to the solidification point.

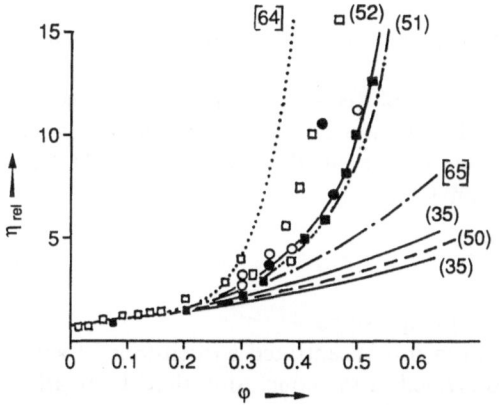

Fig. 2. Concentration dependences of the viscosity of dispersions according to Bedeaux [62]. Numbers on the curves correspond to numbers of equations in the text and also to data in Refs. [64], [65]. Experimental values are given from data in [58], [68]

Figure 2 presents a comparison, published by Bedeaux [62], of the experimental dependences of the viscosity of moderately concentrated and concentrated dispersions (according to the data of Krieger [68] and Russel and de Kruif [58]) with the theoretical data based both on a hydrodynamic approach (Saito (50), Batchelor and Green (35) and on the averaging theories (Lundgren [64], Beenakker [65], Russel (51), Bedeaux (52)). Using a more up-to-date approach it is possible to describe the drastic increase of viscosity in concentrated dispersions more accurately, but, unfortunately, only for a rather narrow range of viscosity increase ($\eta_{rel} < 20$).

The majority of theoretical relations given above depend on a single parameter — the volume fraction of disperse phase particles-φ and apart from this includes only numerical constants. Careful analysis, however, shows that the equations of

Vand (36), Mooney (37), Bedeaux (58), Ladd (61) and also equations (29) of the Huggins's type implicity include the maximum particle packing factor φ_{max} (for example, $[\eta] \sim 1/\varphi_{max}$). In an explicit form the parameter φ_{max} enters into most of the equations corresponding to so-called cellular models. A model of this type was proposed for the first time by Simha [67]. It involved two interacting particles of radius a, and the high concentration of the surrounding dispersion was taken into account by the introduction of the spherical surroundings of radius b around each particle. An equation for the dependence of viscosity on concentration corresponding to this model is as follows:

$$\eta_{rel} = 1 + 5/2\lambda(\gamma)\,\varphi\,, \tag{62}$$

where $\lambda = \dfrac{4(1 - \gamma^7)}{4(1 + \gamma^{10}) - 25\gamma^3(1 - \gamma^4) + 42\gamma^5}$, and $\gamma = a/b$. The packing density of particles defines the value of parameter f: $(\gamma^3 = \varphi/f^3)$. For dispersions of maximum concentration, Eq. (62) has a simpler form:

$$\lim \eta_{rel}|_{\varphi \to \varphi_{max}} = \frac{54}{f^3} \frac{\varphi^2}{(1 - \varphi/\varphi_{max})^3}\,. \tag{62a}$$

For hexagonal packing of particles in dispersion f = 1.81, and for simple cubic packing f = 1.61.

Modifying Mooney's equation (37) in terms of Ree and Eyring's theory used for particle doublets, Krieger and Dougherty [68] arrived at an equation similar in form to several phenomenological equations presented above (17, 20, 21), but, unlike those equations, containing parameter φ_{max}:

$$1/\eta_{rel} = (1 - \varphi/\varphi_{max})^{[\eta]\,\varphi_{max}}\,, \tag{63}$$

where $[\eta]$ is the characteristic viscosity of dispersion.

Analysing the rate of energy dissipation in concentrated suspensions of solid spheres, Frankel and Acrivos [69] also arrived at the conclusion that the relative viscosity of suspensions is a function of the relation a/h:

$$\eta_{rel}|_{h/a \to 0} \sim C(a/h)\,, \tag{64}$$

where a is the radius of particles, h is the gap between them, C is a numerical constant. In the case of cubic packing of particles the value of this relation can be calculated:

$$\frac{a}{h} = \frac{1}{2}\left[\frac{(\varphi/\varphi_{max})^{1/3}}{1 - (\varphi/\varphi_{max})^{1/3}}\right]\,. \tag{64a}$$

Sengun and Probstein [70] drew the same conclusion when they analyzed the flow of concentrated monomodal dispersions of solid particles at extremely high shear rates. They obtained an expression valid in the range of diluted and concentrated dispersions:

$$\eta_{rel} = 1 + C\left(\frac{3\pi}{8}\right)\left(\frac{\beta}{\beta + 1}\right)$$
$$\times \left[\frac{3 + 4.5\beta + \beta^2}{\beta + 1} - 3\left(1 + \frac{1}{\beta}\right)\ln(1 + \beta)\right], \qquad (65)$$

where β is the relation between the size of particles and the distance between them according to Frankel and Acrivos $\left(\beta = \dfrac{2a}{h}\right)$. For dilut dispersions expression (65) is of the form:

$$\eta_{rel} = 1 + C\left(\frac{3\pi}{32}\right)\left(\frac{1}{\varphi_{max}^{4/3}}\right)\varphi^{4/3} \qquad (65a)$$

and for concentrated dispersions they obtained:

$$\eta_{rel} = C\left(\frac{3\pi}{8}\right)\beta. \qquad (65b)$$

The approach developed within the scope of the energy dissipation principle led Quemada [71] to a simpler expression for determination of the viscosity of maximum concentrated dispersions which is based on the relationship φ/φ_{max}:

$$1/\eta_{rel}|_{\varphi/\varphi_{max} \to 1} = (1 - \varphi/\varphi_{max})^2. \qquad (66)$$

In diluted solutions Quemada's theory leads to the following equation:

$$1/\eta_{rel} = 1 - K_0\varphi, \qquad (66a)$$

where K_0 is a coefficient, which for spherical particles is equal to 2.5, which in form coincides with equation (47).

Application of Simha's cellular model, made more complex by Happel [72], gave a somewhat different expression for the interaction factor between particles in concentrated dispersion:

$$\eta_{rel} = 1 + 5.5\psi\varphi, \qquad (67)$$

where

$$\psi = \frac{4\gamma^7 + 10 - 84/11\gamma^2}{10(1 - \gamma^{10}) - 25\gamma^3(1 - \gamma^4)}.$$

Equation (64) was later refined by Shishkin [73], who took into account the presence of a hierarchy of particle assemblies in concentrated dispersions. Shishkin introduced into Eq. (67) the distribution function (F_c) over the volume of the system of unstable cells whose particles are capable of changing their relative position during flow. He arrived at the following equation:

$$\eta_{rel} = \frac{1 - F_c + F_c^2}{F_c} (1 + 5.5\psi\varphi).$$

(68)

The fluctuation model of disperse media, developed by Shishkin is based on a statistical estimation of coordination numbers in a chaotic system of rigid particles. Taking into account the experimental data of Oda [74], Shishkin introduced the following hierarchy of particles with respect to their degrees of freedom. Particles with the coordination number $n < 4$ are free (unstable) according to Bernal [75] and Scott [76]. Particles with $n > 4$ are not free. Free particles exist in a system consisting of spheres at $0.30 < \varphi < \varphi_{max}$ and have an approximately constant number of contacts $n \simeq 3$. Particles that are not free appear at $\varphi > 0.3$–0.35 and have the coordination number $n = 4$–12. The distribution function of free particles throughout the whole system volume is described by the following:

$$F_c = \begin{cases} \exp\left[- \dfrac{\varphi/\varphi_{max} - \varphi_0/\varphi_{max}}{1 - (\varphi/\varphi_{max})^{5/2}} \right], & \text{if } \varphi/\varphi_{max} \leqq \varphi_d/\varphi_{max} \\ \dfrac{1 - \varphi/\varphi_{max}}{1 - \varphi_d/\varphi_{max}} \exp\left[- \dfrac{\varphi_d/\varphi_{max} - \varphi_0/\varphi_{max}}{1 - (\varphi_d/\varphi_{max})^{5/2}} \right], & \text{if } \varphi/\varphi_{max} > \varphi_d/\varphi_{max} \end{cases},$$

(68a)

where φ is the volume fraction of particles in the dispersion, φ_0 is the concentration at the moment "not free" (stable) particles appear in the system, φ_d is the maximum

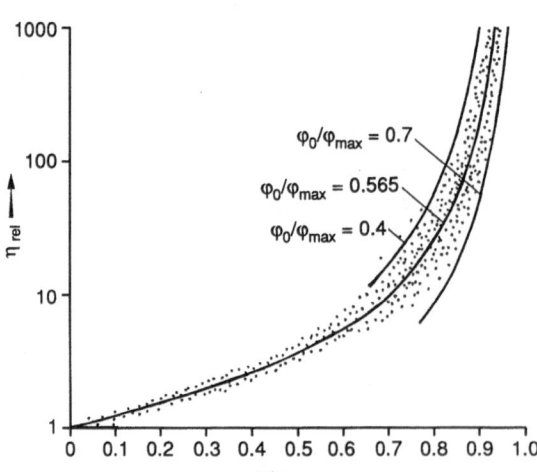

Fig. 3. Dependence of viscosity on the reduced concentration of dispersions according to Shishkin [73]. Theoretical curves correspond to Eq. (68)

concentration of particles in the system which as yet does not exhibit dilatancy, φ_{max} is the maximum packing density of particles. Shishkin pointed out that in the process of deformation, a concentrated dispersion undergoes self-organization, acquiring an optimum structure.

Figure 3 shows the dependence of the relative viscosity of mono- and polyfraction dispersions on the reduced value of their volume concentration (φ/φ_{max}) according to Shishkin's data [73]. The theoretical dependence corresponds to equation (68) at $\varphi_0/\varphi_{max} = 0.565$. The boundaries of theoretical dependences in the range $0.4 < \varphi_0/\varphi_{max} < 0.7$ are given. By means of Eq. (68), which is based on Shishkin's fluctuation model, it is possible to describe the change in relative viscosity of dispersions up to the range $\varphi \rightarrow \varphi_{max}$ within the limits of the change of η_{rel} by three decades of an order.

2.3 The Transition Region from Fluid to Solid-like Disperse Systems

Quemada [71], Bedeaux [49], Mellema-Willemse [60], Beenakker [65], de Gennes [77] considered in detail the reasons of a drastic (infinite) increase in viscosity of concentrated dispersions at $\varphi \rightarrow \varphi_{max}$ even for systems containing spherical particles. Quemada [71] and de Gennes [77] suggested that this concentration range should be considered in terms of the percolation theory developed by Broadbent and Hammersley [78] and Kirkpatrick [79]. Quemada points out that the concentration range corresponding to $\eta \rightarrow \infty$ exists both for monomodal (with $\varphi_{max} = 0.5-0.6$) and for polyfractional dispersions, with $\varphi_{max} \rightarrow 1$. He suggests that transition to the region of infinitely increasing viscosity should be considered as a phase transition from a deformable to a nondeformable system (infinite cluster). De Gennes [77] also believes that at sufficiently high φ, ordering of disperse phase particles becomes comparable with a transition of percolation type. According to Bedeaux [49], at $\varphi \rightarrow \varphi_{max}$ the viscosity of such a system passes through a high maximum but does not attain infinite values. It is not possible to obtain direct evidence owing to the difficulties involved in experimental substantiation of that or any other hypothesis.

To illustrate, solution of problems of this kind in the percolation theory, Quemada [71] cites the analysis of the dependence of electrical conductivity on concentration for a mixture of conducting and nonconducting spherical particles carried out by Fitzpatrick [80] or Clerk [81]. According to Clerk, this dependence is described by an equation similar to Eq. (66) or other similar formulas:

$$G \sim (p - p_{cr})^t \tag{69}$$

where P_{cr} is the critical content of conducting particles in the mixture (percolation threshold) and $t = 1.6-2.0$.

It is interesting that similar conclusions are drawn by the authors who consider the viscosity of liquids and cluster systems in terms of the free volume theories. These will be briefly reviewed in the next chapter.

3 Viscosity of Molecular Systems and Free Volume Theories

3.1 Phenomenological Approaches

Batschinski [82] was the first to notice a relationship between the viscosity of liquids and the volume not occupied by liquid molecules (free volume). He formulated the relation between the viscosity of a large number of liquids and solutions and their specific volume as follows:

$$\eta = \frac{C}{V - \omega}, \qquad (70)$$

where C is a constant, V is the specific volume of liquid and ω is the volume of its molecules; or in a different form:

$$1/\eta = K(V - \omega). \qquad (70a)$$

The variable ω was defined by Batschinski as the specific volume of the liquid with infinite viscosity. It is interesting that the relationship ω/V_{cr} (where V_{cr} is the critical liquid volume) proved to be constant and equal to approximately 0.307 for all the liquids investigated by Batschinski.

Macleod [83] also noted the dependence of the viscosity of liquids on the free space volume, and defined this dependence with the following expression:

$$\eta_t x_t^A = const, \qquad (71)$$

where η_t is the viscosity of liquid at temperature t, x_t is the free space volume related to the volume of molecules at the same temperature. For normal liquids A = 1, for associated liquids A > 1.

Investigating the viscosity of a homological series of liquid normal paraffins, Doolittle [84] pointed out that the direct relationship between viscosity ("resistance to flow") and free volume ("relative volume of molecules per unit free space") is an intuitive hypothesis and the experimental dependence is described better by a logarithmic equation

$$\eta = A \exp\left[\frac{B}{(V_f/V_0)}\right], \qquad (72)$$

where B is a constant and V_f/V_0 is the relative free volume for a given substance, which at constant pressure depends only on temperature, $V_f = V - V_0$, where V is the volume of 1 g liquid at a given temperature, and V_0 is the volume of the same liquid extrapolated to absolute zero temperature.

3.2 Theoretical Models

As pointed out by Doolittle, the relationship between the viscosity of liquids and
their free volume remained for a long time only an intuitive hypothesis though it
described quite well numerous experimental results. A theoretical approach to the
solution of the problem of the relationship between the viscosity of liquid and its
free volume was generalized for the first time by Eyring [85] in terms of the absolute
reaction rates theory. The formulas obtained by Eyring pointed to a qualitative
relationship between viscosity and the ratio of the volume occupied by liquid
molecules C to the volume occupied by "holes" through which molecules "jump"
to the neighboring position:

$$\eta \sim \frac{C}{V - b}, \tag{73}$$

where $(V - b)$ is the volume of "holes". It should be noted that, as pointed out
by Eyring, the free volume depends on the type of packing of molecules in it and
can be obtained by geometric calculation, using rigid spheres as models for liquid
molecules. Such an approach was developed by Frenkel [86] in the well-known
kinetic theory of liquids.

Cohen and Turnbull [87] generalized somewhat the theoretical concepts of the
relationship between diffusion and self-diffusion of liquids modelled by assemblies
of rigid spheres and obtained on the basis of the theories of Frenkel and Eyring,
Fox and Flory [88] and Williams, Landell and Ferry [89] the equation:

$$D = ga^*U \exp\left[-\gamma V^*/V_f\right], \tag{74}$$

where D is the diffusion constant, a^* is the molecular diameter, U is the kinetic
velocity of molecules, g is the geometric factor, γ is the overlap factor $(1/2 - 1)$,
γV^* is the molar volume. At $V \rightarrow V^* D \rightarrow 0$, hence in the glass transition region,
the viscosity of the liquid tends to infinity.

In a later work Cohen and Turnbull [90] defined the free volume of liquid as
a part of the thermal expansion being freely redistributed throughout the volume
without any change in energy. Macedo and Litovitz [91] pointed out that
Cohen-Turnbull's equation (74) ensues from an already known equation obtained
independently by Fulcher [92] and Tamman [93]:

$$\eta = A \exp\left[\frac{1}{\alpha'(T - T_c)}\right], \tag{75}$$

where A is a constant, α' is the thermal expansion coefficient of the liquid. They
proposed complementing the exponential term of Eq. (74) by the energy activation
barrier value:

$$\eta = A_0 \exp\left(E_v^*/RT + \gamma V^*/V_f\right), \tag{76}$$

where A_0 is a constant and E_v^* is the activation barrier.

The relationship between viscous flow and the free volume of liquids capable of undergoing glass transition was established in the so-called valence configurational theory of Nemilov [94], who obtained this relationship in the general form:

$$\eta = A_T \exp\left[\frac{E^*(T)}{RT}\right], \tag{77}$$

where $A_T = N_A h/v^*$, N_A is the Avogardo number, h is the Planck constant, v^* is the activation volume, E^* is the activation term $E^*(T) = E_0^* + E_c(T)$. The general formula for the dependence obtained by Nemilov is in a complex form.

$$\lg \eta = \lg A_l(T_m)$$

$$+ \left[H\eta^*(T_m) + \frac{Z^*}{P} \int_T^{T_m} \left[\int_T^{T_m} Cp_{conf} \, d \ln T \right] dT \right] \frac{1}{4.57T}, \tag{78}$$

where Z^*/P is the number of molecules or structural units falling within the configurational structure transformation region ($n \sim 10^2$). Cp_{conf} is the molecular configurational heat capacity of liquid. By means of Eq. (78) it is possible to calculate the viscosity values of the liquid at any temperature, even in the supercooling region. For calculations using Eq. (78) it is necessary to know the dependences of the thermodynamic functions of liquid and crystal $H_\eta^*(T_m)$ and $S_\eta^*(T)$ on temperature, the valence structure of liquid, its viscosity at one temperature and the temperature viscosity coefficient. The terms $\ln A_l(T_m)$ $= \ln A_T - \frac{S_\eta^*(T)}{R}$ and $H_\eta^*(T_m)$ are determined graphically from experimental results. The results of calculations by means of Eq. (78), complex as it is, coincide well with the experimental results of viscosity determination when cooling liquids in the range of 15 decimal orders. The concepts of "the configurational structure transformation" region advanced by Nemilov seem to be similar to those of the clusters theory.

The respective survey of the modern free volume theories is borrowed from the monograph by Rostiashvili, Irzhak, and Rozenberg [95].

3.3 The Transition Region from Mobile to Glassy Systems

Cohen and Grest [96] developed the thermodynamic aspect of describing the viscosity of dense liquids and glasses. They distinguished between "solid" and "liquid" cluster cells. This made it possible to use the percolation theory for describing the behavior of such a system [78, 79]. According to the Cohen-Grest theory, each of the atoms of the system corresponds to a cell of volume V formed by the surrounding atoms. The type of cell ("liquid-like" or "solid-like") depends on the sign of the difference ($V - V_{cr}$), where V_{cr} is the critical value.

The free volume ($V_f = V - V_{cr}$) can be freely redistributed throughout the whole bulk of the system, and this redistribution does not involve any energy consumption. However, for this free redistribution V_f to be realized it is necessary that among the nearest neighbors of the liquid-like cell there should be a certain number of cells of the same type. It is assumed that in a cluster formed by liquid-like cells each cell has at least two such neighbors (the coordination number $Z > 1$).

It should be noted that, on one hand, an approach such as this is sufficiently closely related to the fluctuation theory of disperse systems developed in Shishkin's works [73], and on the other hand, it reduces to one of the variants of the flow problems in the percolation theory [78, 79] according to which the probability of the existence of an infinite liquid-like cluster depends on the value of the difference $(P - P_{cr})$, where P_{cr} is the flow threshold. At $P < P_{cr}$, only liquid-like clusters of finite dimensions exist which ensure the glassy state of liquid. It is assumed that at $P > P_{cr}$ and $(P - P_{cr}) \ll 1$ the flow probability is of the following scaling form:

$$P_{cr}(p) = B(p - p_{cr})^{\beta_{cr}}, \qquad (79)$$

where B is a constant, and β_{cr} is the critical exponent.

The authors of the cluster theory draw the conclusion that the theory affords a sufficiently rigorous theoretical derivation of Doolittle's equation (72). Verification of the free volume theory advanced by Cohen and Grest was carried out by Hiwatari using computer simulation [97], showed that glass transition in liquids can really be described in terms of the percolation theory, the value of P_{cr} in this case being close to 0.2. Unlike Cohen and Grest's assumptions, however, this transition is not accompanied by a drastic change in the fluidity of the liquid near P_{cr}.

It is interesting that Eq. (79) obtained in describing the dependence of viscosity of liquids whose molecules are modelled by solid spheres or their assemblies is very similar in its form to Eq. (69) and to equations describing the dependence of viscosity of disperse systems on concentration also modelled by assemblies of solid spheres.

4 The Flow and Deformation of Concentrated Disperse Systems in the Light of Free Volume Concepts

4.1 Comparison of the Flow Theories of Disperse Systems with the Free Volume Theory

Analysis of the dependence of viscosity on the concentration of disperse systems and on the free volume of condensed liquid systems shows that there is a considerable similarity between the concepts based on the description of the properties of these systems. This is evidently explained by the similarity of geometric models describing the behavior of these systems based on the description of the

interactions between regularly or randomly packed solid spheres. It is remarkable
that both the flow theories of disperse systems and their free volume theories in
the range of high concentrations of model spheres are related to one of the
percolation theory parts, according to which the dependence, say, of the electrical
conductivity (G) of a system on the volume fraction of conducting filler, can be
expressed by Eq. (69)). Figure 4 shows schematically the dependence of conductivity on the concentration of conducting particles in the mixture. It is characteristic
of this dependence that the G values other than zero are recorded only when critical
concentration is reached.

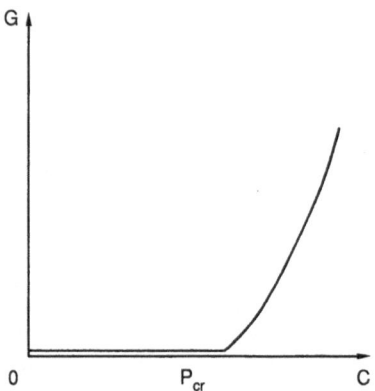

Fig. 4. Scheme of the change in electrical conductivity of a mixture of conducting and nonconducting spheres depending on the conentration of
spheres.
P_{cr} — critical concentration

A similar dependence can be obtained by comparing the fluidity $(1/\eta)$ of a
concentrated disperse system with the volume fraction of the liquid phase in it
$(\varepsilon = 1 - \varphi)$, where φ is the filler content. This dependence is shown on Fig. 5.
As long as the liquid phase volume in a disperse system does not exceed ε_{cr}, i.e.
as long as the filler content in it does not drop below φ_{max}, there is no flow
$(1/\eta = 0)$. Thus the dependence of fluidity on the liquid phase content in dispersion

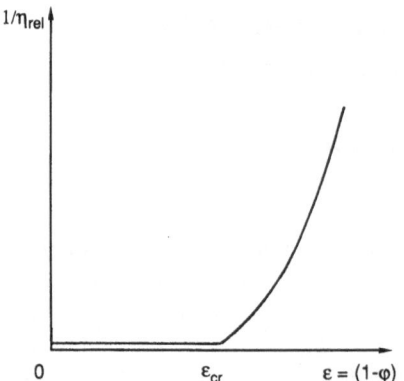

Fig. 5. Scheme of the change in the fluidity of
dispersions with decrease in their concentration. $\varepsilon = (1 - \varphi)$ — dispersion medium con-
tent, ε_{cr} — critical value: $\varepsilon_{cr} = (1 - \varphi_{max})$

can be expressed by the following relation:

$$1/\eta \sim (\varepsilon - \varepsilon_{cr})^t, \tag{80}$$

or in a different form:

$$1/\eta \sim (\varphi_{max} - \varphi)^t \equiv \varphi_f^t, \tag{80a}$$

where

$$\varphi_f = \varphi_{max} - \varphi \tag{80b}$$

and t is the critical exponent. The last mentioned expression is similar to Batschinski's (70) or Macleod's (71) equation. On the other hand, it can be easily transformed to the expression:

$$1/\eta \sim K(1 - \varphi/\varphi_{max})^t, \tag{81}$$

or to the expression:

$$\eta \sim 1/K(1 - \varphi/\varphi_{max})^{-t}, \tag{81a}$$

where K is a constant $(K = \varphi_{max}^t)$, and t is the critical exponent whose value can vary from 1 to 3. Equations (81) and (81a), in their turn, are similar to the theoretical equations obtained by Quemada (66) or Krieger and Dougherty (63) and to the empirical expressions of Landell (20) or Dalla Valle (17). It is evident that the two approaches based on outwardly different promises lead to similar results and, as it seems to us, help one to understand that φ/φ_{max} is not simply a normalized concentration but has a significant physical sense.

If we consider, from the same standpoint, Doolittle's equation (72) whose validity was proved by Cohen and Grest [96], we see that by taking into account the variables introduced into Eq. (80), we can write it as follows:

$$\eta = A \exp\left(\frac{B\varphi}{\varphi_f}\right), \tag{82}$$

where A and B are constants. This equation in a somewhat modified form is as follows:

$$\eta = A \exp\left[\frac{C\varphi}{1 - \varphi/\varphi_{max}}\right], \tag{82a}$$

where C is a constant; it is similar to the equation obtained by Mooney (37). The above equations are, of course, only of a qualitative nature and may, possibly, require theoretical confirmation. The concept of the free volume of disperse systems, however, which takes them into account, to our mind is sufficiently substantiated.

We tried to compare the phenomenological equations (81) and (82) with the experimental results obtained in studies on the slow (creeping) flow of highly-concentrated disperse systems containing a polyfractional filler [98].

4.2 The Relationship Between Mechanical Properties and the Free Volume of Disperse Systems

The viscosity of concentrated dispersions was investigated at room temperature using Volarovich's rotational viscometer with the working module "coaxial cylinders with a hemispherical bottom" [99]. To prevent slippage the surface of the inner cylinder was grooved (grooves-1 mm deep, spaced 2 mm apart). Experiments were carried out at constant torque created by different loads through a block and pulley and transferred to the rotating inner cylinder. It was established by special experiments that for the torque values obtained, beginning with a certain distance from the cylinder surface, the shear stress in dispersion becomes less than its yield point θ. This made it possible to consider the flow of dispersion in the gap as the case of the rotation of the inner cylinder in an infinite volume, and calculating the shear stress (τ) and the shear rate ($\dot{\gamma}$) by the following formulas [100]:

$$\tau = \frac{M}{2\pi R_1^2 L}, \tag{83}$$

$$\dot{\gamma} = 2\omega \frac{d \ln \omega}{d \ln M}, \tag{84}$$

where M is the torque, ω is the angular velocity of the inner cylinder of radius R_1 and length L. The parameters of the working module of the viscometer were: $R_1 = 9$ mm, $L = 50-70$ mm. In calculating the viscosities of dispersions it proved that the contribution of the spherical part of the working module of viscometer to the total shear stress value was not large and did not exceed 5% of the overall stress. Therefore in calculating τ and $\dot{\gamma}$, the parameters of the spherical part of the working module were not taken into consideration.

Experimental studies of the slow flow of highly-concentrated systems were made on polyfractional filler dispersions in glycerol whose viscosity was 1.1 Pas. The parameters of initial filler fractions are listed in Table 1. Of these fractions, mixtures of different composition were made, which served as the disperse phase of the systems being investigated. The composition of mixtures and the concentrations of dispersions are listed in Table 2. The average size of filler particles of coarse and medium fractions was determined by standard sieve analysis and that of fine fraction from its sedimentation rate in media of different viscosity by the instrument "Micronphotosizer."

The values of the maximum packing fraction φ_{max} for each narrow filler fraction were found experimentally by filling the voids in it with an inert low-molecular liquid (water, ethanol).

Table 1. Characteristics of fillers

No. of fractions	Material of filler	Density kg/m^3	Size of particles max	Size of particles min	µm average	φ_{max}	Shape of particles
1.	granite	2650	10000	5000	7500	0.61	irregular
2.	granite	2650	2500	1500	2000	0.53	rhombo-
3.	granite	2650	1500	1000	1250	0.57	hedrical, with
4.	granite	2650	1250	630	940	0.59	anisometry
5.	granite	2650	630	315	470	0.62	of 1.5–2.0
6.	granite	2650	315	140	230	0.62	
7.	granite	2650	200	80	140	0.54	
8.	quartz	2650	320	200	260	0.69	close to spherical
9.	andesite	2700	40	20	30	0.64	irregular

Table 2. Composition of dispersions

Series	Nos	Volumetric composition of dispersions — coarse fraction	medium fraction	fine fraction	dispersion medium	φ_{max}	Free volume φ_f
1	2	3	4	5	6	7	8
0	01–04	—	fraction 8	fraction 9			
	01	—	0.488	0.162	0.350	0.870	0.220
	02	—	0.525	0.175	0.300	0.870	0.170
	03	—	0.562	0.188	0.250	0.870	0.120
	04	—	0.600	0.200	0.200	0.870	0.070
1	10–14	fraction 8	fraction 7	fraction 9			
	10	0.500	0	0.150	0.350	0.843	0.193
	11	0.375	0.125	0.150	0.350	0.798	0.148
	12	0.333	0.167	0.150	0.350	0.771	0.121
	13	0.250	0.250	0.150	0.350	0.736	0.086
	14	0.167	0.333	0.150	0.350	0.699	0.049
2	21–25	fraction 3	fraction 7	fraction 9			
	21	0.375	0.125	0.150	0.350	0.814	0.164
	22	0.333	0.167	0.150	0.350	0.790	0.140
	23	0.250	0.250	0.150	0.350	0.766	0.116
	24	0.167	0.333	0.150	0.350	0.722	0.072
	25	0.125	0.375	0.150	0.350	0.698	0.048
3	31–35	fraction 3	fraction 8	fraction 9			
	31	0.375	0.125	0.150	0.350	0.860	0.210
	32	0.333	0.167	0.150	0.350	0.859	0.209
	33	0.250	0.250	0.150	0.350	0.854	0.204
	34	0.167	0.333	0.150	0.350	0.850	0.200
	35	0.125	0.375	0.150	0.350	0.843	0.193

Table 2. (continued)

Se-ries	Nos	Volumetric composition of dispersions			disper-sion me-dium	φ_{max}	Free volu-me φ_f
		coarse fraction	medium fraction	fine fraction			
1	2	3	4	5	6	7	8
4	41–45	fraction 2	fraction 8	fraction 9			
	41	0.375	0.125	0.150	0.350	0.867	0.217
	42	0.333	0.167	0.150	0.350	0.864	0.214
	43	0.250	0.250	0.150	0.350	0.858	0.208
	44	0.167	0.333	0.150	0.350	0.854	0.204
	45	0.125	0.375	0.150	0.350	0.851	0.201
5	51–53	fraction 3	fraction 7	fraction 9			
	51	0.428	0.143	0.129	0.300	0.835	0.135
	52	0.381	0.190	0.129	0.300	0.805	0.105
	53	0.285	0.286	0.129	0.300	0.725	0.025
6	61–64	fraction 2	fraction 8	fraction 9			
	61	0.428	0.143	0.129	0.300	0.793	0.093
	62	0.381	0.190	0.129	0.300	0.792	0.092
	63	0.285	0.286	0.129	0.300	0.788	0.088
	64	0.190	0.381	0.129	0.300	0.786	0.086
7	71–75	fraction 2	fraction 3	fraction 9			
	71	0.434	0.217	0.090	0.250	0.890	0.140
	72	0.480	0.240	0.079	0.200	0.860	0.160
	73	0.527	0.264	0.059	0.150	0.838	−0.012
	74	0.574	0.287	0.040	0.100	0.820	−0.080
	75	0.592	0.300	0.032	0.080	0.810	−0.110
8	81–82	fraction 1	fraction 5	fraction 9			
	81	0.539	0.210	0.110	0.141	0.830	0.081*
	82	0.506	0.197	0.130	0.167	0.830	0.127*
	83–84	fraction 1	fraction 6	fraction 9			
	83	0.570	0.192	0.104	0.134	0.824	0.162*
	84	0.482	0.195	0.141	0.182	0.838	0.162*
	85	fraction 1	fraction 4	fraction 6	fraction 9		
	85	0.512	0.147	0.080	0.114	0.147	0.854	0.113*

* data refer to $\varphi_{f, n-1}$

Of particular interest was the question of the width of the concentration range in which it is possible to study experimentally the flow of dispersions with a polyfractional filler prone to settling. On one side this range is bounded by the value of φ_{max}, on the other side the concentration range in which dispersions of constant structure exist is bounded by their stability to settling. The settling process of disperse phase in the dispersions being investigated was complicated by their being polyfractional. The coarsest filler fraction settled faster than all other particles, i.e. its settling rate limits the stability of the system as a whole. Collisions with particles of finer fractions, however, hinder settling of coarse particles. We tried to take this effect into account, assuming as a first approximation that quantitatively it shows itself as an increase in viscosity and density of the complex dispersion medium due to the presence of finer fractions. A dispersion was assumed to be stable if during the test time (10^2 s) it settled no more than by 1% along the height of the working module of the instrument. The required stability of dispersion based on glycerol proved to be reached at the total volume fraction of filler $\varphi \geq 0.65$. Special experiments on determination of the sedimentation rate of filler in dispersion confirmed this value. It should be noted that in the dispersions under investigation, settling of the coarsest particles in a high-viscosity loaded medium takes a very peculiar form. Particles of coarse filler fractions forming the greatest part of the polyfractional system can rest on one another and on finer particles and, presumably, form a labile skeleton taking up the whole mixture volume and breaking reversibly upon its deformation. The higher the deformation intensity, the greater is the extent of breaking of the initial labile mixture skeleton. Due to the change of the mixture composition, i.e. the change of the packing fraction of polyfractional filler φ_{max} and its volume fraction, fluid nonsettling compositions were obtained with the free volume value of $0.05 \leq \varphi_f \leq 0.2$.

Figure 6 (a–e) shows the flow curves of highly concentrated disperse systems based on glycerol. Analysis of these curves shows that the range of stresses and shear rates in which flow is possible is very limited. At shear stresses below the yield point there is no flow. A significant increase in shear stress leads to the wall effect, i.e. to loss of contact between the rotating cylinder of the viscometer and the dispersion undergoing deformation. This effect is usually connected with decreased disperse phase concentration in the dispersion layer adjoining the rotating viscometer cylinder due to the hydrostatic pressure normal to the shear surface which arises during the rotation of moving particles [101]. Another possible reason of this effect is the impossibility for the structure of highly-loaded dispersions to undergo reorganization at an arbitrary rate equal to that of its deformation. If for dispersions with a relatively low viscosity and large interparticle distance, deformation (and structure reorganization) can occur at a fairly high rate without appearance of wall effects, then for dispersions with a limiting viscosity (at disperse phase content close to φ_{max}) structure reorganization upon deformation of the system, even at extremely low rates, can lead to its breaking contact with the shear surface. Possibly, such highly-concentrated dispersions with rough an-isometric particles cannot flow at all, and under deformation lose their continuity.

In view of the specific features of the flow of highly-concentrated polyfractional dispersions with particles of irregular shape, mentioned above, sufficiently full flow curves could be obtained only for a relatively limited set of dispersions listed in Table 2 (series 0–6). The flow curves obtained show a marked pseudoplasticity, the reason of which seems to be dynamic reorganization of their structure during flow. Deformation results in the breaking of the original random structure of dispersions; it undergoes ordering which tends to arrange particles in a regular packing form, in which fine particles are localized among larger particles uniformly distributed in the mixture. In that case their mutual influence becomes minimum and the effective viscosity of the dispersion decreases. Possibly, the preferential

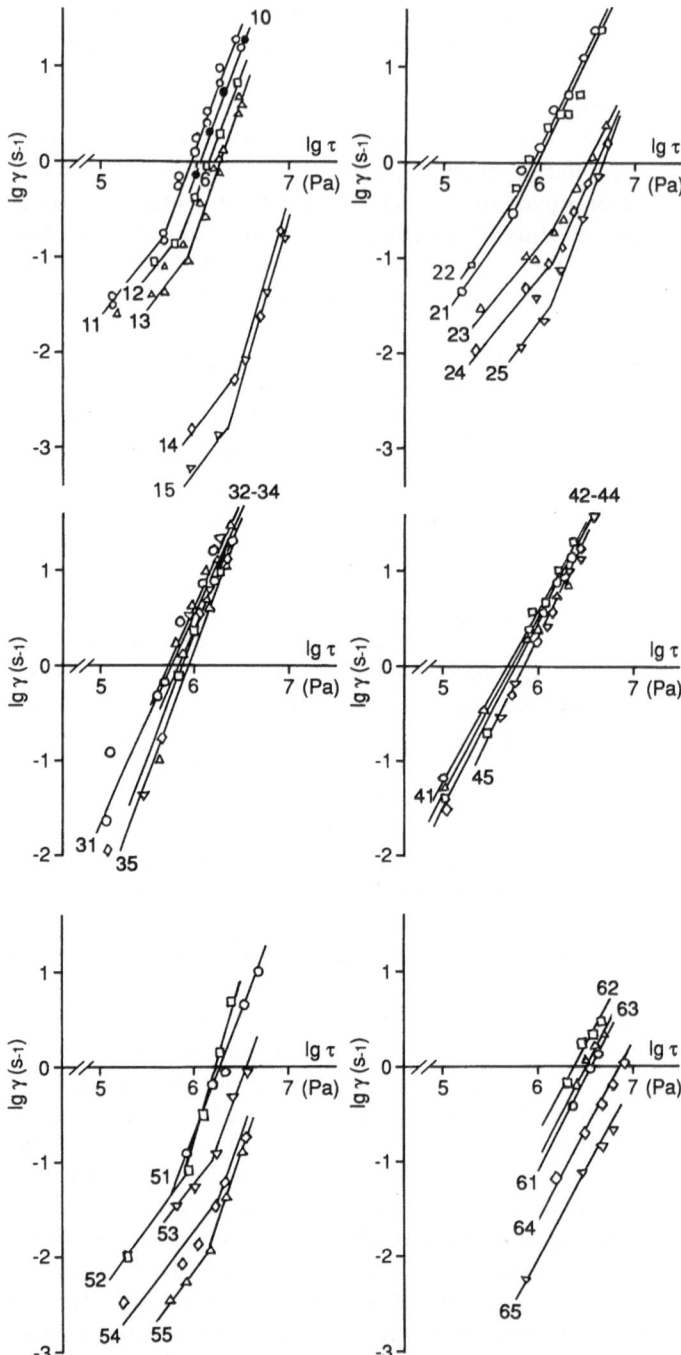

Fig. 6. Flow curves of concentrated dispersions of polyfractional fillers (series 1–6) in glycerol. Numbers on the curves correspond to the numbers of the mixtures in Table 2

orientation of filler particles of irregular shape also makes a certain contribution to the decrease in viscosity of dispersions, though their anisotropy is not great.

On the basis of the measured values of the viscosity of concentrated dispersions in view of Eq. (81) and (82) (see the beginning of this section), we tried to estimate the dependence of relative viscosity of dispersions on their "free volume." The results are given in Figs. 7 and 8.

In an analytical form the dependence shown on Fig. 7 is:

$$\lg \eta_{rel} = 3.16 - 3.16 \lg (\varphi_{max} - \varphi). \tag{85}$$

The same equation written in the form similar to Batschinski's formula (70) is:

$$\eta_{rel}^{-0.316} = 0.1(\varphi_{max} - \varphi), \tag{85a}$$

or in the form similar to Landell's formula:

$$\eta_{rel} = 1.45 \cdot 10^3 (\varphi_{max} - \varphi)^{-3.16}. \tag{85b}$$

The high value of the proportionality coefficient between the relative viscosity coefficient of dispersions and their "free volume", in our opinion, is due to the extremely high contribution of the surface friction of contacting particles in such highly-concentrated dispersions.

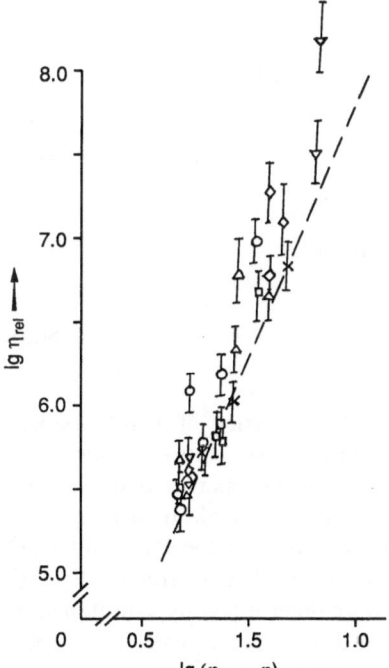

Fig. 7. Dependence of the logarithm of relative viscosity of dispersions on the logarithm of their free volume (according to data in Fig. 6)

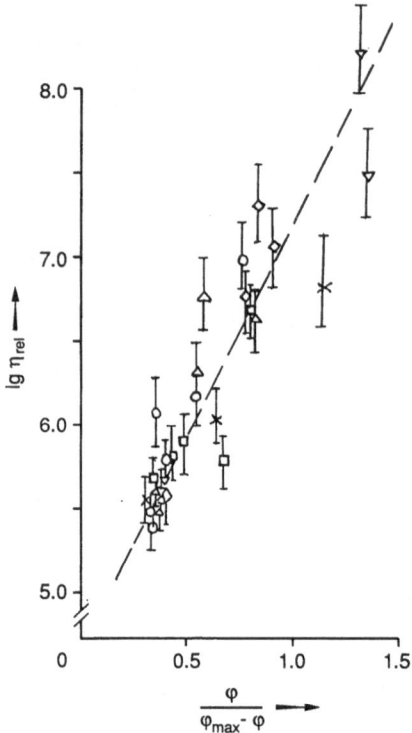

Fig. 8. Dependence of the logarithm of relative viscosity of dispersions on the value of the function $\varphi/(\varphi_{max} - \varphi)$ (according to data in Fig. 6)

The analytical expression of the dependence shown on Fig. 8 is of the form:

$$\lg \eta_{rel} = 4.80 + 0.24\,\frac{\varphi}{\varphi_{max} - \varphi}, \tag{86}$$

or in the form corresponding to Doolittle's formula (72):

$$\eta_{rel} = 6 \cdot 10^4 \exp\left(\frac{0.55\varphi}{\varphi_{max} - \varphi}\right). \tag{86a}$$

Not only viscosity but also other mechanical properties of highly-loaded composites in the unhardened state depend on their free volume value. The experiments which enabled estimation of some mechanical properties of unhardened composites with different polyfractional filler content (breaking stress, shear modulus, work of breaking) were carried out with the use of a special installation – a modification of Rehbinder-Weiler's plastometer. The characteristics of highly-loaded composites (see Table 2, series 7) were determined by simultaneous measurement of stresses and deformations during drawing of a grooved cylinder from a cylindrical casing with grooved walls, the gap between them being filled

with packed composition material. The binding agent for composites in this series
was liquid furan resin with a viscosity of 0.04 Pa · s. The binding agent content and
the polyfractional filler composition in composite varied within wide limits (see
Table 2, series 7) including both the range of $0 < \varphi_f \leqq 0.2$ and the range $\varphi > \varphi_{max}$
relating to composites containing air inclusions which cannot be removed even
at maximum packing. The values of mechanical characteristics of highly-loaded
composites with different free volumes are given on Fig. 9. With increasing φ_f, all
the mechanical characteristics of the composites investigated decrease. Increase
of the volume fraction of filler above its maximum packing (the case $\varphi_f < 0$) also
leads to a decrease of mechanical characteristics due to the weakening effect of
air voids in the mixture. The mixtures studied had maximum resistance to
deformation at $\varphi_f \rightarrow 0(\varphi \rightarrow \varphi_{max})$.

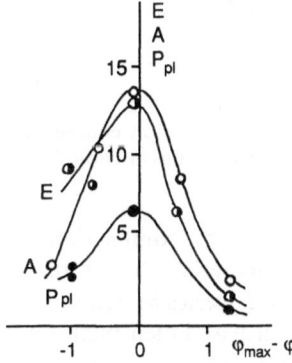

Fig. 9. Dependence of mechanical properties of highly-
loaded composites (series 7, Table 2) on their free volume.
P_{pl} — breaking stress, MPa $\times 10^{-2}$.
A — work of breaking, N · m;
E — shear modulus, MPa

4.3 The Mobility of Disperse System Particles and Its Change with the Free Volume

The approach based on analysis of the free volume of disperse systems proved to
be useful not only in calculating the viscosity of dispersions flowing without loss
of continuity, but also in estimating the mobility of particles in highly-concentrated
composites containing solid, liquid and gaseous (air) phases, in particular in
investigating their capability of compaction under vibratory action [102]. Increase
in the density of such highly-loaded composites during compaction is described
by curves which cannot be approximated by simple equations. Some curves of
this kind are shown in Fig. 10. The complex nature of the curves seems to be due
to the fact that several processes take place during compaction: fast processes of
breaking of the primary labile skeleton of dispersion particles and their repacking,
a slower mutual approach of particles and their orientation into the most
convenient position and, finally, the slowest process of the exit of air inclusions
and compression of the mixture [103]. The rate of all these processes, except the
last step, is determined by the mobility of dispersion particles.

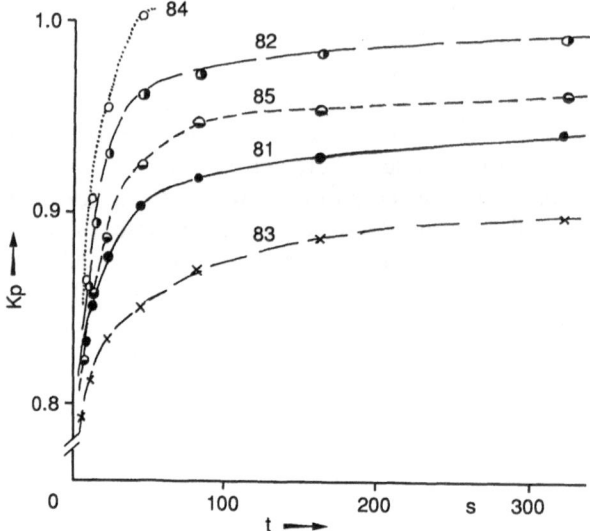

Fig. 10. Dependence of the packing factor of highly-loaded coarse-dispersion composites on vibration time. Numbers on the curves correspond to numbers of composites in Table 2

Experimental studies on the changes in density upon vibrocompaction of highly-loaded composites were carried out for coarse materials based on a polyfractional filler, whose composition is given in Table 2 (series 8) containing as liquid phase low-molecular liquids (acetone, water, glycerol and their mixtures), typical thermoreactive resins based on furan, epoxy, polyester oligomer. The viscosity of dispersion media ranged within 0.0003–45 Pa · s.

The kinetics of vibrocompaction of highly-loaded disperse systems with different polyfractional filler content was studied with the use of vibration equipment which allowed variation in a relatively wide range of vibration frequency (10–200 Hz) and the amplitude of vibrations (0.05–5 mm), measuring at regular intervals the composition density (the mixture column height in the mold). To normalize the density values in data processing the value of the compaction coefficient K_p was used which is equal to the ratio of the varying composite density and its maximum possible value. Using semilogarithmic coordinates (K_p – lg t), the change of $K_p(t)$ in time was expressed by two intersecting straight line segments. The nature of the experimental curves for the variation of the compaction degree with time plotted on these coordinates is illustrated in Fig. 11. We think that each linear section of the dependences corresponds to a definite step of the compaction process which has its own mechanism, and thus the whole compaction process proceeds in two steps. In the first step the rate of compaction of composites is many times higher than in the second step and therefore the fast step is of greatest interest for investigation. The segment of the straight line characterizing the fast compaction step can be described by the following equation:

$$K_p = A + B \lg t ,$$

(87)

where A and B are coefficients, constant for definite compositions of mixtures and for definite compaction parameters (amplitude and vibration frequency, external pressure on the mixture). Investigation of the parameters determining the behavior of mixtures during compaction showed that at constant intensity of vibrations for binders of the same viscosity, change in the polyfractional filler composition and content in dispersion does not affect the value of coefficient A, but alters the value of coefficient B in Eq. (87). More detailed experiments on a large number of different highly-loaded composites containing from 60 to 95 vol.% of polyfractional fillers of different composition showed that coefficient B depends only on the structure of composites, and the vibrocompaction parameters and the liquid phase viscosity practically do not affect it.

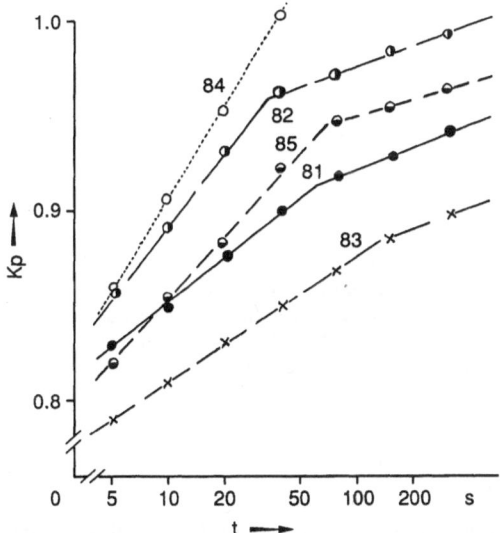

Fig. 11. Dependence of the packing factor of highly-loaded coarse-dispersion composites on vibration time on semilogarithmic coordinates. Numbers on the curves correspond to the numbers of the composites

In a search for the defining structural parameter of a composite, the free volume of disperse system proved to be the most sound one from the physical standpoint. Presumably, for disperse systems the free volume is a measure of the mobility of filler particles, just as for liquids it is a measure of the mobility of molecules. But as applied to highly-loaded coarse systems of the type "solid particles — liquid — gas" this notion requires a certain correction. In characterizing the structure of such specific systems as highly-loaded coarse composites, it should be noted that to prevent their settling and separation into layers under the action of vibration, the concentration of the finest filler fraction with the largest specific surface in dispersion medium should be the maximum possible. Because of this and also because of the small size of particles (20–40 μm), the fine fraction suspended in the dispersion medium practically does not participate in the formation of the composite skeleton, which consists of coarser particles. Therefore for the case given, Eq. (80b) should be used in the following form:

$$\varphi_{f,n-1} = |\varphi_{max} - \varphi|_{n-1},\tag{88}$$

where subscript $n-1$ means that in considering the behavior of the mixture, all filler fractions, except the finest, are taken into account.

Analysis of experimental kinetic dependences of vibrocompaction of heavily-loaded composites showed that the structure-sensitive coefficient B in Eq. (87) correlates best with the free volume of dispersions in compact state $\varphi_{f,n-1}$. This dependence is shown in Fig. 12 (the correlation coefficient is 0.951).

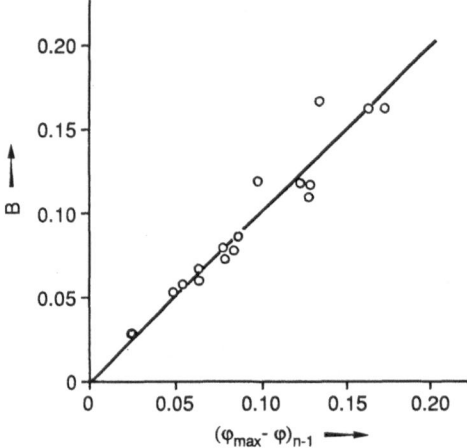

Fig. 12. Correlation between the value of the structure-sensitive coefficient B from Eq. (87) and the free volume of composites

The concept of the free volume of disperse systems can also be correlated with the change in the structure of the composite of the type "solid particles — liquid — gas" during its compaction. In that case the value of the maximum packing fraction of filler φ_{max} in Eq. (80b) remains valid also for systems containing air inclusions, and instead of the value of the volume fraction of filler, characteristic for a "solid particles — liquid" dispersion-φ, its value in an uncompacted system "solid particles — liquid — gas" should be substituted. This value can be calculated as follows: the ratio of concentrations C_{s-1-g}/C_{s-1} to the first approximation can be substituted by the ratio of the densities of uncompacted and compacted composites, i.e. by parameter K_p. Then Eq. (80b) in view of Eq. (88), for uncompacted composites acquires the form:

$$\varphi'_{f,n-1}(t) = \varphi_{max,n-1} - \varphi_{n-1} \cdot K_p,\tag{89}$$

where $\varphi'_{f,n-1}(t)$ is the current free volume of composite with account for packing of coarse and medium filler fractions.

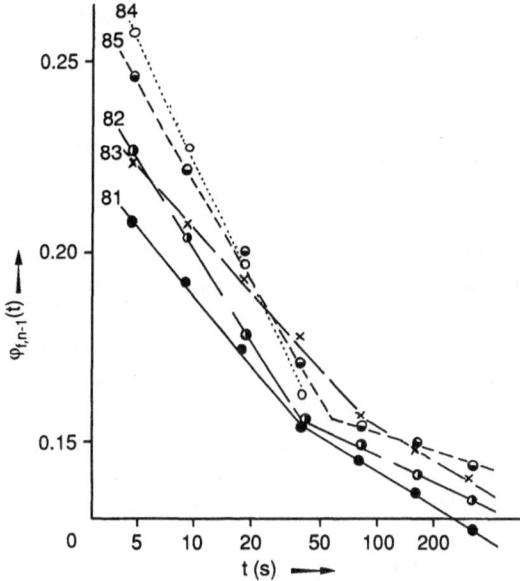

Fig. 13 Dependence of $\varphi'_{f,n-1}(t)$ on vibration time. Numbers on the curves correspond to numbers of composites

Analysis of the dependence (89) (shown in Fig. 13 for different composites) showed that this dependence like a similar dependence shown in Fig. 11 is expressed by two linear segments corresponding to the two compaction steps. It should be stressed that at the breaking point, evidently related to the change in the mobility of dispersion particles, the value of $\varphi'_{f,n-1}(t)$ proved practically the same for mixtures with different polyfractional filler content and composition. This very interesting fact can be explained by drawing an analogy between the kinetic processes characteristic for molecular and disperse systems. In the first compaction step particles of a heavily-loaded disperse system are mobile, in the second step they are hardly moving. The transition point from one compaction step to another is the boundary between two states of the disperse system. If the value of $\varphi'_{f,n-1}(t)$ drops below the critical value (0.15–0.17), particles of coarse and medium dispersion fractions lose their mobility. This phenomenon can be compared with glass transition in polymer systems. Transition from the highly-elastic state characterized by the mobile state of macromolecules to the glassy (low-mobility) state also occurs at a constant value of polymer free volume. In our case the fast vibrocompaction step can be compared with the "fluid" state of polymer and the slow compaction step with the glassy state; the transition from one step to the other is comparable to glass transition.

The relationship established between the change of the mobility of highly-concentrated composite particles and the change of the "free volume" in it warrants certain technological predictions. Thus, for mixtures whose composition complies with the condition $\varphi'_{f,n-1} \geqq 0.15$–$0.17$ rapid processing by moulding without injection is possible or by vibromoulding with a minimum intensity of vibrational action. For mixtures whose composition complies with the condition $\varphi'_{f,n-1} < 0.15$

moulding proceeds in two steps, i.e. it is more time- and energy consuming, though the formulation of mixtures in this case is more economical.

4.4 Determination of the Maximum Packing Fraction of Polydisperse Fillers and Optimization of the Composition of Composites

For reliable application of the free volume concept of disperse systems one must have dependable methods of determination of the maximum packing fraction of the filler φ_{max}. Unfortunately, the possibility of a reliable theoretical calculation of its value, even for narrow filler fractions, seems to be problematic since there are practically no methods available for calculations for filler particles of arbitrary shape. The most reliable data are those obtained by computer simulation of the maximum packing fraction for spherical particles which give the value $\varphi_{max} = 0.625–0.630$. Even relatively narrow fractions of real fillers are, as a rule, charcterized by φ_{max} values lying in a wider range. The deviations observed in the direction of lesser φ_{max} values are usually associated with possible particle aggregation, so that they are probable for fractions of small particle size. Deviations of φ_{max} towards larger values are usually observed for fillers whose particle shape is nearly cubic. At present the most reliable method of determination of φ_{max} for narrow fractions is experiment.

In passing to polyfractional fillers of different composition, the labor intensity of experimental determination of φ_{max} increases many times as it is necessary to test all possible combinations of narrow fractions of different particle size and different content of particles in the mixture. However, it is possible to determine φ_{max} for polyfractional materials with very wide, or discontinuous, particle size distributions by numerical calculation. Methods based on geometric modeling of the packing of spherical particles of different size, unfortunately, are not applicable to the description of the properties of real fillers. A fairly simple method of numerical calculation of φ_{max} for mixtures of different fractions of arbitrarily shaped particles is known, however, this is based on a combination of experimental determination of packing factors (or to be more precise, of porosity coefficients. $e = (1 - \varphi_{max})/\varphi_{max}$) for individual fractions, and a set of equations permitting estimation of their change with changing mixture composition [104]. The set of equations for calculation of the change in the porosity coefficient of a two-fraction mixture under the assumption that the coarser (or the finer) fraction forms the mixture skeleton, is of the form:

$$\left. \begin{array}{l} x_1e_1 + x_2(K''_{21} - K''_{21}e_2 - 1) = e^I \\ x_1e_1K'_{21} + x_2e_2 \qquad\qquad = e^{II} \\ x_1 + x_2 \qquad\qquad\qquad\quad = 1 \end{array} \right\}, \tag{90}$$

where x_1, e_1 and x_2, e_2 are the volume fractions and porosity coefficients of each particle fraction; e^I and e^{II} are the porosity coefficients for the mixture of fractions. Coefficients K'_{21} and K''_{21} are determined by the ratio of the sizes of fractions being mixed ($\psi = \bar{d}_2/\bar{d}_1$):

$$\left.\begin{array}{l} K'_{21} = \dfrac{\psi(1 + 2\psi)}{\psi(1 + 2\psi) + (1 - \psi)^2} \\[4mm] K''_{21} = \dfrac{\psi^2(3 + \psi)}{\psi^2(3 + \psi) + (1 - \psi)^3} \end{array}\right\} . \qquad (90\,a)$$

At the optimum point corresponding to the maximum packing fraction, $e^{I} = e^{II}$, since the porosity value at this point does not depend on whether fine or coarse particles form the mixture skeleton. The Equation (90) permit calculation of the porosity values for a mixture of fractions of any composition. A similar procedure can be used for calculating the porosity of mixtures of three or more fractions. A laboratory check of these Equations (90) confirmed good agreement between the calculation forecast and the experimental determination of porosity coefficients and φ_{max} values for a wide variety of combinations of narrow filler fractions.

It should be noted that a similar approach can be used for an optimum composition of a composite based on a polyfractional filler with a maximum possible packing fraction. One should remember, however, in that case that for a correct choice of the composite composition it is necessary to take into account the interactions existing at the mineral filler — liquid binder interface and leading to formation on the filler surface of the so-called boundary layers [105]. A comparatively simple method for a rough estimation of these interactions for highly-loaded composites is to be found in [106].

4.5 Structurally Similar Approaches

The analog method, based on the structural similarity of the systems being described or the models adopted for their description which we used in working out the concept of the "free volume of disperse systems" has lately undergone further development. For illustration, we can quote the recently published work of Evans and Gibson [107] in which they discuss on an equivalent basis the dynamics of liquid-crystalline systems and composites with short fibers, comparing the free volume of rigid-chain molecules with the "free volume" arising due to rotation of short fibers in a liquid polymeric or thermoreactive matrix. The above authors point out that the similarity of the principles underlying the models can be used for solving problems encountered in analysis of the behavior of systems differing significantly in their scale.

A similar approach has also been developed by Susteric [108], who compares the behavior during low-amplitude deformation of rubbers, loaded with aggregated carbon black, with the visco-elastic behavior of macromolecules undergoing high-frequency deformation. The specific features of the breaking of carbon black aggregates defined by the deformation amplitude of loaded rubbers are described by the above author by a mathematical model developed for the description of the dynamic, visco-elastic behavior of polymer molecules. This approach revealed some new aspects in the behavior of loaded systems in spite of a great difference in the scale of the phenomena being compared.

The above examples show that the use of the analog method offers promise, at any rate, for describing various deformation properties of filler-containing systems. The main positive feature of this approach is the possibility of clarifying the physical picture of phenomena occurring in imperfectly understood systems on the basis of known models as well as the possibility of using certain theoretical relations, derived for description of phenomena of one scale, for phenomena on a different scale.

5 Conclusion

The material presented in this survey contains a great number of formulas describing the dependence of viscosity on the composition of disperse systems. Availability of so many different formulas is primarily due to the fact that, in the large concentration range, quantitative description of the viscosity of suspension presents great difficulties. For this viscosity range it is difficult to develop an adequate model which would permit the universal calculation of viscosity or fluidity of such systems. The idea of taking into account the limiting volume fraction of filler proposed by Mooney proved to be very fruitful. But actually, taking account of the limiting volume fraction amounted only to the statement of the fact that at packing fractions above φ_{max} the system losses its ability to flow. Therefore, the concentration range in which the system is able to flow decreases by the value of φ_{max}.

In some more recent theories, it was suggested that the free volume of the system could really be taken into account in describing the rheological properties of suspensions. This direction, however, is not sufficiently developed.

The central point of the present survey is an attempt to show a complete analogy between the free volume of suspensions and that of molecular systems. It is characteristic that the limiting volume fraction of spherical filler particles leaves in the system another 25–40% of "unoccupied" volume. Precisely the same "unoccupied" volume exists in molecular systems if we liken them to a volume filled with spheres whose radii are calculated taking into account the Lennard-Jones potential.

The glass transition in a molecular system sets in at the moment the free volume (about 2.5% of the free volume in glass transition of polymers) disappears. Loss of fluidity ("glass transition") in suspensions occurs when the free volume of suspensions tends to zero.

The experimental data obtained by us, presented in this survey and compared with literature data show that the free volume is, presumably, a structure parameter characterizing most reliably the properties of highly-loaded suspensions. Using this parameter is most beneficial when we have to deal with a polydisperse suspension and it is difficult to describe it by means of a model.

It should be pointed out that it is in this work that the striking similarity of the structure and relationships between the structure and properties of molecular systems and suspensions is demonstrated for the first time. Actually, calculating the suspension concentration from the limiting volume fraction of filler bears a close analogy to the Williams-Landell-Ferry principle where temperature increase calculated from the glass transition temperature leads to universal and equal increase of free volume for all systems. Our data show that in suspensions, increase of free volume with decreasing system concentration leads to equal increase of flow in different systems if the free volume increase is calculated from the maximum packing fraction.

References

1. Einstein A (1905) Ann Phys 17: 549; (1906) Ann Phys 19: 289; (1911) Ann Phys 34: 591; (1920) Kolloid-Z 27: 137
2. Hatschek E (1913) Trans Faraday Soc 9: 80
3. Baker F (1913) J Chem Soc 103: 1653
4. Hess WR (1920) Kolloid-Z 27: 1
5. Orr C, Blocker HG (1955) J Colloid Sci 10: 24
6. Arrhenius S (1887) J Phys Chem 1: 285
7. Weltmann RN, Green H (1943) J Appl Phys 14: 569
8. Cit by [16]
9. Nicodemo L, Nicolais L (1974) Polymer 15: 589; (1974) also J Appl Polym Sci 18: 2809
10. Tangsathitkulachai S, Austin LG (1988) Powder Technology 56: 293
11. Ting AP, Luebbers RH (1957) AlChE Journal 3: 111
12. Robinson JV (1949) J Phys & Coll Chem 53: 1042; (1951) J Phys & Coll Chem 55: 455; (1951) also Trans Soc Rheol 1: 15
13. Mori Y, Ototake N (1956) Kagaku Kogaku — Chem Eng (Japan) 20: 488
14. Trawinski H (1953) Chem Ing Tech 25: 229
15. Leva M (1959) Fluidization, New York
16. Rutgers R Jr (1962) Rheol Acta 2: 305
17. Dalla Valle J, Orr C (1954) Chem Eng Progr Symp 50: 29; also Zenz FA, Othmer DF (1961) Fluidisation and fluid particles systems, NY p 451
18. Hawksley PGW (1951) Some aspects of fluid flow, ch 7, Inst Phys Arnold, London
19. Kynch GJ (1959) Nature 184: 1311
20. Landell RF, Moser BG, Bauman A (1965) Proc 4th Internat Congr Rheol Part 2, Lee EH (ed) Intersci NY, p 663
21. Pliskin I, Tokita N (1972) J Appl Polym Sci 16: 473
22. Eilers H (1941) Kolloid-Z, Z-Polym 97: 313; (1943) Kolloid-Z, Z-Polym 102: 154
23. Fedors RF (1975) Polymer 16: 305; (1974) also J Colloid & Interface Sci 46: 545
24. Chong JS, Cristiansen EB, Baer AD (1971) J Appl Polym Sci 15: 2007
25. Pal R, Rhodes E (1989) J Rheol 33: 1021
26. Johnston CW, Brower CH (1970) SPE Journal 26: 31. Cit by [27]
27. Williams, GE, Bergen JT, Poechlin GW (1979) J Rheol 23: 591
28. Eirich FR (ed) (1956) Rheology. Theory and applications. Acad Press Intersci Publishers. NY, vol 1, ch 14
29. Happel J, Brenner H (1965) Low Reynolds number hydrodynamics with special application to particulate media, Prentice-Hall, Englewood Cliffs NJ
30. Kunitz M (1926) J Gener Physiol 9: 715
31. Brinkman HC (1947) Appl Sci Res A1: 27; (1947) Appl Sci Res A2: 81; (1947) Appl Sci Res A3: 333; (1952) also J Chem Phys 20: 571
32. Heller W (1954) J Colloid Sci 2: 547
33. Guth E, Simha R (1936) Kolloid-Z 74: 266
34. Vand V (1948) J Phys & Colloid Chem 52: 277; (1948) J Phys & Colloid Chem 52: 299

35. Kynch G (1954) J Appl Phys 3: 55; (1956) also Proc Roy Soc London A237: 90
36. Lewis TB, Nielsen LE (1968) Trans Soc Rheol 12: 421
37. Petersen JM, Fixman M (1963) J Chem Phys 39: 2516
38. Bedeaux D, Kapral R, Mazur P (1977) Physica 88A: 88
39. Batchelor, GK, Green JT (1972) J Fluid Mech 56: 401; (1974) also Ann Rev Fluid Mech 6: 227
40. Mooney M (1951) J Colloid Sci 6: 162
41. Brodnyan JG (1959) Trans Soc Rheol 3: 61
42. Simha R (1940) J Phys Chem 44: 25; (1945) also J Chem Phys 13: 188
43. Kuhn W, Kuhn H (1945) Helv Chim Acta 28: 97; also Kuhn W, Kuhn H, Buchner P (1951) Ergeb exact Naturw 25: 1
44. Kunnen J (1984) Rheol Acta 23: 424
45. Thomas DG (1965) J Colloid Sci 20: 267
46. Roscoe R (1949) J Phys Chem 53: 1042; (1952) also Brit J Appl Phys 3: 267
47. Gillespie T (1960) J Colloid Sci 15: 219; (1963) J Colloid Sci 18: 32; (1960) also J Polym Sci 46: 383; Gillespie T (1963) Emulsion Rheology; Sherman PA (ed) (1963) Pergamon Press, NY p 115
48. de Bruijn H (1951) Disc Faraday Soc 11: 86
49. Bedeaux D (1983) Physica 121A: 345
50. Budianski B (1965) J Mech & Phys Solids 13: 223. Cit by [52]
51. Hill R (1965) J Mech & Phys Solids 13: 213. Cit by [52]
52. Christensen, RM (1979) Mechanics of composite materials. Wiley Intersci NY ch 2
53. Lee DI (1969) Trans Soc Rheol 13: 273; (1970) also J Paint Technol 42: 550
54. Ishai O, Cohen LJ (1967) Int J Mech Sci 9: 539; (1967) Int J Mech Sci 9: 605
55. Narkis M, Nicolais L, Joseph E (1978) J Appl Polym Sci 22: 2391
56. Saito N (1950) J Phys Soc Japan 5: 4; (1950) J Phys Soc Japan 6: 162; (1952) J Phys Soc Japan 7: 447
57. Nagatani T (1979) J Phys Soc Japan 47: 320
58. Wagner AJ, Russel WB (1989) Physica 155A: 475; also de Kruiff CG, van Iersel EMF, Vzij A, Russel WB (1985) J Chem Phys 83: 4717; Russel WB, Gast AP (1986) J Chem Phys 84: 1815
59. Taylor GI (1932) Proc Roy Soc London A138: 41; (1934) Proc Roy Soc London A146: 501
60. Mellema J, Willemse MWM (1983) Physica 122A: 286
61. Felderhof BU (1976) Physica 82A: 596; (1976) Physica 82A: 611
62. Bedeaux D (1987) J Colloid & Interface Sci 118: 80
63. Cichocki B, Felderhof BU, Schmitz R (1989) Physica 154A: 233
64. Lundgren TS (1972) J Fluid Mech 51: 273
65. Beenakker CJW (1984) Physica 128A: 48
66. Ladd AJC (1988) J Chem Phys 88: 5051; (1989) J Chem Phys 90: 1149
67. Simha R (1952) J Appl Phys 23: 1020
68. Krieger IM, Dougherty, TJ (1959) Trans Soc Rheol 3: 137; also Krieger IM (1972) Adv Colloid & Interface Sci 3: 111; also Krieger IM, Maron SH (1960) Reology. Theory and application. Ed by Eirich FR vol 3 Acad Press NY
69. Frankel NA, Acrivos A (1967) Chem Eng Sci 22: 847
70. Sengun MZ, Probstein RE (1989) PCH (Phys-Chem Hydrodynamics) 11: 229
71. Quemada D (1977) Rheol Acta 16: 82
72. Happel J (1957) J Appl Phys 28: 1288
73. Shishkin VA (1980) Proc 2nd Symp. Theory and mech process of polym mater perm p 6 (in russian); Shishkin VA (1984) The structure-mechanical study of the composite materials and constructions. USSR Acad Ural Sci Center Perm p 109 (in Russian)
74. Oda M (1977) Solids foundation 17: 29
75. Bernal ID, Mason I (1960) Nature 188: 910
76. Scott D (1962) Nature 194: 956
77. de Gennes PG (1979) J Physique 40: 783

78. Broadbent SR, Hammersley, JM (1957) Proc Camb Phylos Soc (Math & Phys Sci) 53: 629; (1957) Proc Camb Phylos Soc (Math & Phys Sci) 53: 642
79. Shante VKS, Kirkpatrick S (1971) Adv in Phys 20: 325; also Kirkpatrick S (1973) Rev Mod Phys 45: 574
80. Fitzpatrick JP, Malt RB, Spaepen F (1974) Phys Lett 47A: 207
81. Clerc J, Girand G, Roussenq J (1975) CR Acad Sc Paris 281B: 227
82. Batschinski A (1913) Z für Phys Chem 84: 643
83. Macleod DB (1923) Trans Faraday Soc 19: 6; (1923) Trans Faraday Soc 19: 17
84. Doolittle AK (1951) J Appl Phys 22: 1471
85. Eyring H (1936) J Chem Phys 4: 283; also Hirschfelder J, Stevenson D, Eyring H (1937) J Chem Phys 5: 896; (1937) also J Phys Chem 41: 249
86. Frenkel YaI (1959) Kinetic theory of liquids. USSR Moscow – Leningrad USSR Acad Sci
87. Cohen MH, Turnbull D (1959) J Chem Phys 31: 1164
88. Fox TG, Flory PJ (1950) J Appl Phys 21: 581; (1951) also J Phys Chem 55: 221; (1954) also J Polym Sci 14: 315
89. Williams MF, Landell RF, Ferry JD (1955) J Amer Chem Soc 77: 3701
90. Turnbull D, Cohen MH (1961) J Chem Phys 34: 120
91. Macedo PB, Litovitz TA (1965) J Chem Phys 42: 245
92. Fulcher GS (1925) J Amer Ceram Soc 8: 339
93. Tamman G, Hesse W (1926) Z Anorg Allgem Chem 156: 245
94. Nemilov SV (1976) Fizika i khimiya stekla USSR (Phys & Chem of Glass) 2: 97; (1976) Fizika i khimiya stekla USSR (Phys & Chem of Glass) 2: 192; (1977) Fizika i khimiya stekla USSR (Phys & Chem of Glass) 3: 423; (1978) Fizika i khimiya stekla USSR (Phys & Chem of Glass) 4: 129 (in Russian)
95. Rostiashvili VG, Irzhak VI, Rosenberg BA (1987) Glass-transition of polymers. USSR Leningrad. Ed Khimiya (in Russian)
96. Cohen MH, Grest S (1979) Phys Rev B20: 1077; Grest S, Cohen MH (1980) Phys Rev B21: 4113
97. Hiwatary Y (1982) J Chem Phys 76: 5502
98. Kandyrin LB, Kuleznev VN, Shcheulova LK (1983) Kolloidnyi zhurnal (USSR) 45: 657
99. Volarovich MP (1944) The low-temperature viscosity of lubricant oils. Moscow Ed USSR Acad Sci (in Russian)
100. Middleman S (1968) The flow of high polymers. Continuum and molecular rheology. Intersci Publ Wiley NY–London–Sydney–Toronto
101. Karnis A, Goldsmith HL, Mason SG (1966) Canad Chem Eng 44: 181; also Shizgal B, Goldsmith HL, Mason SG (1965) Canad Chem Eng 43: 97
102. Kandyrin LB, Kuleznev VN, Vorobyev LR, Belnik PR (1987) Mekhanika kompozitnykh materialov LatSSR (Mech of Compos Mater) No 4: 719 (in Russian)
103. Uryev NB (1980) High-concentrated disperse systems. USSR Moscow. Ed Khimiya (in Russian)
104. Wieckowski A, Strek F (1966) Chemia stosowana (Poland) 1B: 95; (1966) Chemia stosowana (Poland) 3B: 431
105. Kuleznev VN, Markhasin IL, Kondrashov OF, Chernin YeI, Kandyrin LB, Fuchs GI, Grinberg SM (1980) Kolloidnyi zhurnal (USSR) 42: 49
106. Kuleznev VN, Kandyrin LB (1989) Makromol Chem Macromol Symp 28: 267
107. Evans KE, Gibson AG (1986) Compos Sci & Technol 25: 149
108. Susteric Z (1989) Makromol Chem Macromol Symp 23: 329

Editor S. Edwards
Received November 14, 1990

Glassy State Relaxation and Deformation in Polymers

T. S. Chow
Xerox Webster Research Center, 800 Phillips Road, 114-39D,
Webster, NY 14580, USA

An overview of a nonequilibrium glass theory is presented to describe the structural relaxation and deformation kinetics of polymeric glasses, compatible blends, and particulate composites. The glassy state relaxation is a result of the local configurational rearrangements of molecular segments, and the dynamics of holes (free volumes) provide a quantitative description of the segmental mobility. On the basis of the dynamics of hole motion, a unified physical picture has emerged which enables us to discuss the structure relaxation, physical aging, and glassy state deformation. The links between the bulk and shear relaxations, the change in deformation from linear to nonlinear viscoelastic responses, and the nonlinear viscoelastic nature of plastic deformation are discussed. Theoretical expressions are presented for the determination of the PVT (pressure-volume-temperature) behavior, for the elucidation of the equilibrium and nonequilibrium nature of the glass transition, for the calculation of viscoelastic response, and for the prediction of yield behavior and stress-strain relationships of these polymeric systems.

Abbreviations

Latin symbols:

a	macroscopic timescale (shift factor)
A	parameter measures volume interaction
d	fractal dimension
D	local diffusion constant
e_{ij}	strain tensor
E	relaxation modulus
f	hole (free volume) fraction
ΔH	activation energy
k	Boltzmann constant
n	number of holes
n_x	number of polymer molecules
N	total number of lattice sites
p	pressure
q	cooling rate (<0)
q_v	wave number on fractal lattice
$\mathbf{Q}$	wave vector of the fluctuation
$\mathbf{r}$	spatial vector
R	diffusion length
t	time
T	temperature
T_r	reference temperature
T_g	glass transition temperature
v	lattice volume
V	total volume
ΔW	external work acting on the lattice
x	number of monomer segments/polymer

Greek symbols:

α $= \varepsilon f / kT^2$

β stretched exponent

δ nonequilibrium hole fraction

ε_h $= \varepsilon$ is the hole energy

ε_f flex energy

η $= E/2(1 + \theta)$ is the shear modulus

θ Poisson ratio

$\varkappa$ $= E/3(1-2\theta)$ is the bulk modulus

μ physical aging rate

ν fractal exponent

ϱ state density

σ_{ij} stress tensor

σ_y yield stress

τ relaxation time

φ volume fraction in blends or composites

χ interaction parameter

Ψ relaxation function

ω angular frequency

Ω_{ij} activation volume tensor

1 Introduction

There is a strong dependence on temperature and time of the properties of solid polymers [1–3] compared to those of other materials such as metal and ceramics. The strong dependence of properties on temperature and timescale (deformation rate, aging rate, cooling rate, etc.) is a result of the gradual approach of the nonequilibrium glassy state to its equilibrium. The study of volume relaxation [4–8] below the glass transition temperature has indeed revealed that glassy states undergo the slow change processes. The structure relaxation, physical aging, and glassy state deformation of amorphous polymers have been extensively investigated and reported in the literature [1–14]. It has been recognized that both viscoelastic response and plastic yield represent the inherent irreversible nature of deformed glasses. The physics of glassy polymers is still evolving and the functional relationships between relaxation and deformation have not been firmly established. However, significant progress has been made which can be utilized in solving many problems in this area.

This work sets out to provide a unified account of recent advances in understanding the structural and mechanical properties of polymers in the glassy state. Since all properties vary slowly in space and time, the long term behavior becomes the main concern. The glassy state relaxation is treated as a result of the local configurational rearrangements of molecular segments which is described by the hole motion. We shall start with an overview of recent advances in the nonequilibrium statistical mechanical theory of hole motion. This provides us with a solid foundation for a quantitative description of the structural relaxation, physical aging and deformation kinetics. Extending the concept of hole dynamics in the glassy state, equations are obtained (1) for the determination of the equation of state (PVT behavior) of amorphous polymers in both the equilibrium liquid and nonequilibrium glassy states which elucidates the kinetics of glass transition, (2) for the calculation of viscoelastic response, and (3) for the prediction of yield behavior and stress-strain relationships of these polymeric systems. Without adjusting the input molecular parameters for a given polymeric system, the following effects on the mechanical properties of solid polymers can be predicted: the physical aging, strain rate, temperature, the nonequilibrium polymer-polymer interaction in compatible blends, and the filler concentration in composites. The calculated phenomena are compared with the experimental data.

2 Glass Theory

2.1 Equation of Motion

Consider a lattice consisting of n holes and n_x polymer molecules of x monomer segments each. The total number of lattice sites is written in the form

$$N(t) = n(t) + xn_x \tag{1}$$

where $n \ll N$. It is important to mention that n consists of both equilibrium and nonequilibrium contributions in the glassy state. For temperatures above T_g, the nonequilibrium contributions to n go to zero. The change of $n(t)$ below T_g defines the glassy state. Minimizing the excess Gibbs free energy due to hole introduction with respect to the hole number, the temperature dependence of the equilibrium hole fraction is given by [5]

$$\bar{f}(T) = \frac{\bar{n}}{N} = f_r \exp\left[-\frac{\varepsilon}{k} \left(\frac{1}{T} - \frac{1}{T_r} \right) \right] \tag{2}$$

where ε is the mean energy of hole formation, and the subscript r refers to the condition at $T = T_r$, which is a fixed quantity near T_g. The "bar" refers to the equilibrium as a fully relaxed state with the nonequilibrium hole fraction

$$\delta(T, t) = f(T, t) - \bar{f}(T) \tag{3}$$

equal to zero. Equation (2) reveals that holes are created by raising the temperature and are eliminated by lowering it.

Amorphous solids are not in thermodynamic equilibrium. The departures from equilibrium for holes and bond rotations have been treated as a stochastic process. We have reported that the conformational activation energy controlling the rotational relaxation of bonds is between 1 and 2 orders of magnitude lower than the hole activation energy [12]: Λ(conformer) = 1.94 kcal/mol versus Λ(hole) = 75.3 kcal/mol for poly(vinyl acetate), in the vicinity of the glass transition. As a result, the conformer relaxes much faster than the hole. Since all physical properties of glass vary slowly in time (t), the dominant contribution to the structural relaxation and physical aging in glasses is from the hole. The hole configurational space in a quenched and annealed glass is divided into *regions separated by barriers*. To include the spatial vector ($\mathbf{r}$) into the analysis, $n(t)$-$\bar{n}$ is related to the local excess of hole number density $\delta n(\mathbf{r}, t)$ by integrating over the volume surrounding an individual hole

$$n(t) - \bar{n} = \int \delta n(\mathbf{r}, t) \, d\mathbf{r}$$

Consider a polymer is quenched from liquid to glass where the sample is annealed. During isothermal annealing, the number of holes is close to a conserved quantity. The local excess of number density of the quenched glass relaxes by spreading slowly over the entire region, and is governed by

$$\frac{\partial \delta n(\mathbf{r}, t)}{\partial t} = \int [W(\mathbf{r} \mid \mathbf{r}') \, \delta n(\mathbf{r}', t) - W(\mathbf{r}' \mid \mathbf{r}) \, \delta n(\mathbf{r}, t)] \, d\mathbf{r}' \tag{4}$$

where $W(\mathbf{r} \mid \mathbf{r}')$ is the transition probability per unit time jumping from $\mathbf{r}'$ to $\mathbf{r}$, and the integration is over the space. When the necessary condition of convergence is assumed, Eq. (4) can be rewritten as [12]

$$\frac{\partial \delta n(\mathbf{r}, t)}{\partial t} = \sum_{m=1}^{\infty} \frac{1}{m!} \, (-\nabla)^m \, b_m(\mathbf{r}) \, \delta n(\mathbf{r}, t) \tag{5}$$

where b_m is the mth moment of the transition rate $W(\mathbf{r}' \mid \mathbf{r})$:

$$b_m(\mathbf{r}) = \int (\mathbf{r}' - \mathbf{r})^m \, W(\mathbf{r}' \mid \mathbf{r}) \, d\mathbf{r}'$$

Equation (5) is a partial differential equation of infinite order, and cannot be solved in general. Since all properties of glass vary slowly in space and time, the left-hand side of Eq. (5) can be truncated, the motion of holes in response to molecular fluctuations is then treated as an anomalous diffusion process [12]

$$\left(\frac{\partial}{\partial t} - \nabla \cdot D\nabla \right) \delta n(\mathbf{r}, t) = 0 \tag{6}$$

where $D = b_2/2$. We have assumed here that the system is in a quasi-equilibrium state and in the absence of external field. Equation (6) reveals that the dynamics of holes are diffusive and not vibrational. Let us assume that, initially, δn is non-zero only at $\mathbf{r} = 0$. When D is a constant, the solution of Eq. (6) displays the well-known Gaussian spreading. However, we have a spatial dependent diffusion coefficient. Let us introduce the Fourier transform in space,

$$\delta n(\mathbf{Q}, t) = \int \delta n(\mathbf{r}, t) \, e^{-i\mathbf{Q} \cdot \mathbf{r}} \, d\mathbf{r}$$

Equation (6) can be written in the form:

$$\left(\frac{\partial}{\partial t} - DQ^{2+v} \right) \delta n(\mathbf{Q}, t) = 0 \tag{7}$$

where v produces the fractal dimension d, which defines a self-similar scaling between wave numbers:

$$Q \sim q_v^d, \quad \text{with} \quad d = \frac{2}{2 + v} > 0 \tag{8}$$

Either d or v is an independent exponent. By using Eq. (8), Eq. (7) is transformed to

$$\left(\frac{\partial}{\partial t} - D_v q_v^2 \right) \delta n(q_v, t) = 0 \tag{9}$$

On the fractal lattice, D_v is a constant, and holes exhibit the Gaussian characteristics. The self-similarity of the fractal has dilation symmetry shown in Eq. (8).
Using the Fourier-Laplace transformation in time,

$$\delta n[q_v, \omega] = \int_0^\infty \delta n(q_v, t)\, e^{i\omega t}\, dt$$

we obtain from Eq. (9)

$$\delta[\omega] = \sum_{q_v} \frac{\delta(q_v, t = 0)}{D_v q_v^2 - i\omega} \tag{10}$$

which solves the initial value problem and is normalized by N. Note Eq. (10) is
valid only for small wave number or large wave length. Defining the local relaxation
time

$$\tau_v = \frac{1}{D_v q_v^2} \tag{11}$$

the Fourier-Laplace inversion of Eq. (10) is

$$\delta(t) = \sum_{q_v} \delta(q_v, t = 0)\exp\left[-\frac{t}{\tau_v(q_v)}\right] \equiv \sum_{q_v} \delta(q_v, t) \tag{12}$$

where the summation is carried out over the wave numbers up to a cutoff q_m.
Eq. (10) reveals that $\delta(q_v, t)$ is the solution of

$$\frac{d\delta_i(t)}{dt} = -\frac{\delta_i(t)}{\tau_i} \tag{13}$$

where the subscript i is identified with a particular wave number on the fractal
lattice. When a system is also under a temperature change such as cooling, Eq. (13)
has to be modified [12]

$$\frac{d\delta_i(t)}{dt} = -\frac{\delta_i(t)}{\tau_i} - q\,\frac{\varepsilon_i \bar{f}_i}{kT^2} \qquad (i = 1, ..., L) \tag{14}$$

where $q = dT/dt$, ε_i and $\bar{f}_i$ are defined in the same way as those in Eq. (2).
In addition, $\bar{f} = \Sigma \bar{f}_i$, $\varepsilon = \Sigma \varepsilon_i \bar{f}_i/\bar{f}$, and L is an integer which corresponds to the
maximum wave number q_m. Equation (14) has the form of the KAHR multi-
parameter equation [7], except that it is expressed in terms of microscopic
parameters.

2.2 Nonequilibrium State

For a system started from equilibrium, the solution of Eq. (14) has the form [5]

$$\delta(T, t) = \sum_i \delta_i = -\frac{\varepsilon}{k} \int_0^t \frac{q\bar{f}}{T^2} \Psi(t - t')\, dt' \tag{15}$$

Different paths of time integration describe different thermal history behavior of the glassy state relaxation and recovery kinetics. Some typical examples of thermal history paths for Eq. (15) were illustrated in Ref. [5]. In this review, we consider mainly that the polymer is cooled from liquid to glass, and is then followed by isothermal annealing.

In accordance with the anomalous diffusion on fractal lattice, we expect [12, 15]

$$\langle \Delta r^2 \rangle^{1/2} \equiv R \sim t^\beta, \qquad 0 < \beta \leq 1 \tag{16}$$

When the random process is Gaussian, we have $\beta = 1/2$. As we mentioned earlier, there are regions of size (or length l) separated by energy barriers in an amorphous solid. This size is proportional to the timescale (λ) needed for a hole to penetrate a barrier of height (Λ) [16],

$$l \sim \lambda \sim \exp\left(\frac{\Lambda}{kT}\right) \tag{17}$$

Thus,

$$\frac{R}{l} \sim \frac{t^\beta}{\lambda} = \left(\frac{t}{a}\right)^\beta \quad \text{with} \quad a = \lambda^{1/\beta} \tag{18}$$

During isothermal annealing, the frozen-in structure of quenched glass relaxes, and $\Psi(t)$ can be interpreted as the probability that holes have not reached their equilibrium states. The probability of a hole in the ith wave number state having reached equilibrium in a time interval t is $(n_i/n)\,(R/l) \approx R/Ll$. Thus, we write [12]

$$\Psi(t) \simeq \left(-\frac{R}{Ll}\right)^L \to \exp\left(-\frac{R}{l}\right), \quad \text{as} \quad L \to \infty$$

Combining the above equation with Eq. (18) yields

$$\Psi(t) = \exp\left[-\left(\frac{t}{\tau}\right)^\beta\right], \quad 0 < \beta \leq 1 \tag{19}$$

where $\tau = \tau_r a$. The stretched exponential, Eq. (19), has been successfully used to represent the volume relaxation, viscoelastic response, and physical aging in glasses.

It has the familiar form of the Kohlrausch-Williams-Watts (KWW) equation [17], except that β and τ are not empirical constants here, and they will be discussed in the next two sections.

2.3 Relaxation Time

Several well-known equations are available for interpreting the temperature dependence of viscosity, diffusion coefficient, and other relaxation rates for $T > T_g$. The Doolittle equation [18], the WLF equation [19], the Vogel-Fulcher equation [20], and the Adam-Gibbs equation [21] can be expressed in the same form. They are known to fit well with the relaxation data of liquids in equilibrium. The universal functional form is [20]

$$a(T) \sim \exp\left[\frac{J}{T - T_2}\right], \quad \text{for} \quad T > T_g \tag{20}$$

Here J is a constant, and $T_2 \simeq T_g - 50$ K. The above equation is not valid for $T < T_g$ and has an apparent singularity in τ as $T \to T_2$. This basically prevents us from following this line of thought in order to determine the low temperature structural relaxation and physical aging in glassy polymers.

In our lattice model, each lattice site occupies a single lattice cell of volume v. In view of the cooperative nature of the hole motion, the barrier energy is treated by a mean field average [28], and is related to the Gibbs free energy per molecule (Δg) in a system restrained to single occupancy of cells by

$$\Lambda = \frac{N}{n} \Delta g = \frac{N}{n} (\varepsilon + pv - T\Delta s) \tag{21}$$

where p is chosen to be zero in most of our discussion, and Δs is the configuration entropy per molecule. We have

$$\Delta s = k \ln\left(\frac{n}{\bar{n}}\right) = k \ln\left(1 + \frac{\delta}{\bar{f}}\right) \tag{22}$$

Combining Eqs. (2), (18), (21) and (22) yields

$$a(T, \delta) = \left(\frac{\bar{f} + \delta}{f_r}\right)^{-\frac{1}{\beta(\bar{f} + \delta)}} \tag{23}$$

At $T = T_r$, we have $\delta = 0$, $\bar{f} = f_r$, and $\tau = \tau_r$ which require $a = 1$. In practice, the shift factor is measured in logarithmic scale. Equation (23) can be written approximately as

$$\ln a(T, \delta) \simeq \frac{1}{\beta}\left(\frac{1}{\bar{f} + \delta} - \frac{1}{f_r}\right), \quad \text{as} \quad \frac{\bar{f} + \delta}{f_r} \to 1 \tag{24a}$$

Eq. (24a) may be called the "generalized" Doolittle equation. When $T \geq T_g$, we have $\delta = 0$ and Eq. (24a) deduces to the exact form of Doolittle's equation [18].

In the vicinity of T_g, Eq. (24a) can also be written in the form [5, 12]

$$\ln a(T, \delta) \simeq - \frac{\alpha_r(T - T_r) + \delta}{\beta f_r^2}, \qquad \alpha_r = \varepsilon f_r / k T_r^2 \qquad (24b)$$

These relaxation time equations together with Eqs. (2), (15), and (19) can be utilized in analyzing the experimental measurements of volume relaxation and recovery, of linear and nonlinear viscoelastic relaxations, and of yield behavior and stress-strain relationships.

2.4 Relaxation Spectrum

The parameter β should be a function of local interaction which is related to the non-zero ν mentioned in Eqs. (7) and (8). Let us consider the hole density-density correlation function and write the Green's function

$$G(\mathbf{r}, t) = \frac{\langle \delta n(\mathbf{r}, t) \, \delta n(0, 0) \rangle}{\langle \delta n^2 \rangle} \qquad (25)$$

The angular brackets denote an equilibrium ensemble average. $G(\mathbf{r}, t)$ is invariant under translations of $\mathbf{r}$ and t, and vanishes when $\mathbf{r}$ and/or t are very large. Following the same procedure led to Eq. (10), except by using two-sided Fourier time transform, we obtain

$$G(\omega) = \sum_{q_v} \frac{1}{D_v q_v^2 - i\omega} = \int_0^{q_m} \frac{\varrho(q_v) \, dq_v}{D_v q_v^2 - i\omega} \qquad (26)$$

where ϱ is the state density. The number of modes per unit length along the hole path with wave number between Q and $Q + dQ$ can be expressed in terms of the number of modes on the fractal lattice by using Eq. (8):

$$\frac{dQ}{2\pi} \sim \frac{d}{2\pi} q_v^{d-1} \, dq_v$$

Using Eq. (11), we obtain

$$\varrho(q_v) \, dq_v \sim q_v^{d-1} dq_v \sim \tau_v^{-d/2} d\tau_v \qquad (27)$$

Which, due to its diffusive nature, has different time dependence than that of vibrational fractons [22]. Substituting Eq. (27) into Eq. (26) leads to the asymptotic solution:

$$G(\omega) \sim \int_{\tau}^{\infty} \frac{\tau_v^{-d/2} d\tau_v}{1 - i\omega\tau_v} \approx \int_{\tau}^{\infty} \frac{\tau_v^{-d/2} d\tau_v}{-i\omega\tau_v} = -\frac{2}{d} \frac{\tau^{-d/2}}{i\omega}, \quad \text{for} \quad \omega\tau_v \gg 1$$

(28)

where

$$\tau = \frac{1}{D_v q_m^2}$$

(29)

is the macroscopic relaxation time. The change in the state of glass during isothermal annealing is accompanied by dissipation (absorption) of energy, which is related to the density fluctuations of hole from its equilibrium value. In accordance with the method of the generalized susceptibility [23], the viscoelastic loss modulus (E''), which measures the energy dissipation, is determined from Eq. (28) as

$$E''(\omega) \sim \text{Im}G(\omega) \sim \tau^{-d/2}$$

(30)

On the other hand, when Eq. (19) is used, we have derived another asymptotic expression [24] of the loss modulus

$$E''(\omega) \sim (\omega\tau)^{-\beta} \quad \text{for} \quad \omega\tau \gg 1$$

(31)

Comparing Eqs. (30) and (31) yields

$$\beta = \frac{d}{2} = \frac{1}{2 + v}$$

(32)

By looking at Eqs. (27) and (28), Eq. (32) confirms the customary way of relating β of the stretched exponential function, Eq. (19), to the relaxation time spectrum. The glassy state relaxation is dominated by the part of the spectrum having longer relaxation times. The fractal dynamics of holes are diffusive, and the diffusivity depends strongly on the tenuous structure in fractal lattices. v is the exponent in the power-law relationship between local diffusivity and diffusion length:

$$D = \frac{R^2}{2t} \sim R^{-v}$$

(33)

The physics behind the divergence of the diffusivity in Eq. (33) at $R = 0$ for $v > 0$ was revealed when the fractal dimension had been introduced in Eq. (7) together with a spatial scale transformation, Eq. (8). For linear polymers, the diffusion coefficient is spatially independent and $\beta = 0.5$. In crosslinked polymers, the connectivity of hole motion decreases significantly because holes encounter many dead ends and the diffusion coefficient decreases rapidly with distance. Matching the motion of holes and the chain motions in terms of the segmental mobility of an ideal phantom network [25] yields $\nu = 4$ and $\beta = 1/6 \simeq 0.17$. We shall see later that these theoretical predictions compare well with the $\beta = 0.48$ and 0.19 for linear and crosslinked polymers, respectively, obtained by fitting experimental data.

3 Glass Transition

3.1 Role of Chain Conformation

Two different views have been adopted in the molecular interpretation of the glass transition. One view considers conditions when relaxation processes occur so slowly that the glass transition can be treated as time independent phenomenon.

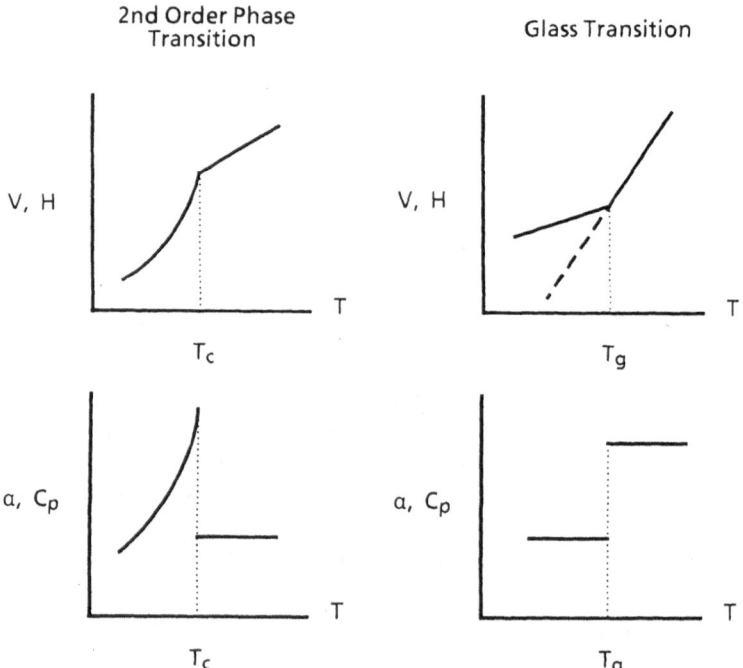

Fig. 1. Second order phase transition versus the vitrification process. They have different temperature dependence of the volume (V) and enthalpy (H), and their derivatives α (thermal expansion) and C_p (specific heat)

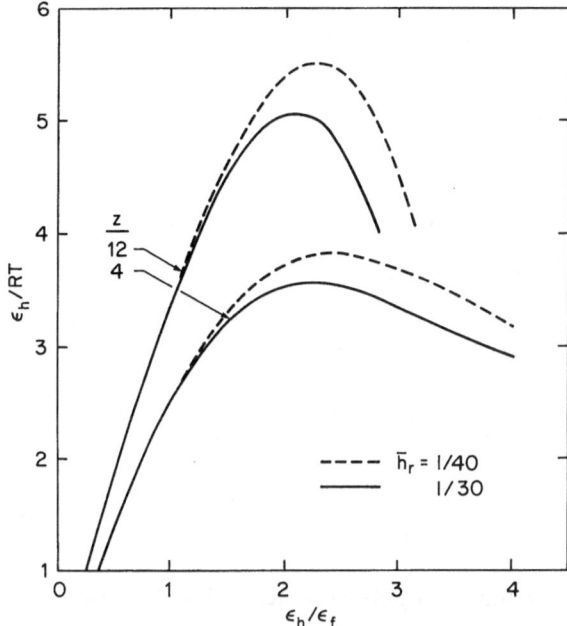

Fig. 2. Maxima of the curves defining the equilibrium glass transition temperature. It is treated as a thermodynamic anomaly at which the most stable hole configuration is reached under the close packing of holes and flex bonds. R is the gas constant, z is the lattice coordinate number, and $h_r(=f_r)$ is the hole fraction at T_r [11]

The other view is directed at the nonequilibrium character of structural relaxation. We will discuss both approaches. The Gibbs-DiMarzio (GD) theory [26] is perhaps the most successful equilibrium theory which takes into account both the conformational (ε_f) and hole (ε_h) energies. The flex energy ε_f is the difference between high-energy gauche and low-energy trans state. In the GD theory, a second order phase transition is identified with the glass transition. In Fig. 1 different temperature dependence of the volume (V) and enthalpy (H), and their derivatives α (thermal expansion) and C_p (specific heat) is shown for second-order phase transition and for the glass solidification. We have replaced the idea of the second-order phase transformation postulated in the GD theory, yet retained its successful features in a new approach. The equilibrium freezing temperature (T_r) is treated as a thermodynamic anomaly at which the most stable hole configuration is reached under the close packing of holes and flex bonds. The maxima of the calculated curves in Fig. 2 define T_r. Table 1 shows the relationships between ε_f, ε_h, and T_r, which remain unchanged in describing other physical properties, of some linear polymers. From Fig. 2 and Table 1, one may write approximately that [11]

$$\frac{\varepsilon_h}{kT_r} \simeq \frac{2.15\varepsilon_f}{kT_r} = \gamma \tag{34}$$

where γ is a function of the lattice coordinate number [11]. The study reveals that the ratio of hole and flex energies is close to a constant. This supports the notion that the conformational theory is experimentally equivalent to the hole theory in the molecular interpretation of T_r.

Table 1. Flex and hole energies, and their ratios of Poly(vinyl acetate) (PVAc), Poly(vinyl chloride) (PVC), and Polystyrene

	PVAc	PVC	PS
T_r, K	308	360	371
ε_h, kcal/mol	2.51	2.92	3.58
ε_h/kT_r	4.11	4.09	4.86
ε_f/kT_r	1.90	1.88	2.36
$\varepsilon_h/\varepsilon_f$	2.16	2.18	2.06
ε_f, kcal/mol	1.16	1.34	1.74

3.2 PVT Behavior

When an amorphous melt is cooled isobarically from liquid to solid, the total volume is [27, 28]

$$V = vN = v[n(t) + xn_x] = V_0 \frac{1 + \delta(t)}{1 - \bar{f}} \tag{35}$$

where $V_0 = vxn_x$ is the occupied volume. For poly(vinyl acetate) (PVAc), the input parameters are

$$\varepsilon = 2.51 \text{ kcal/mol}, \qquad f_r = 0.0336; \qquad \beta = 0.48,$$
$$\tau_r = 25 \text{ min} \tag{36}$$

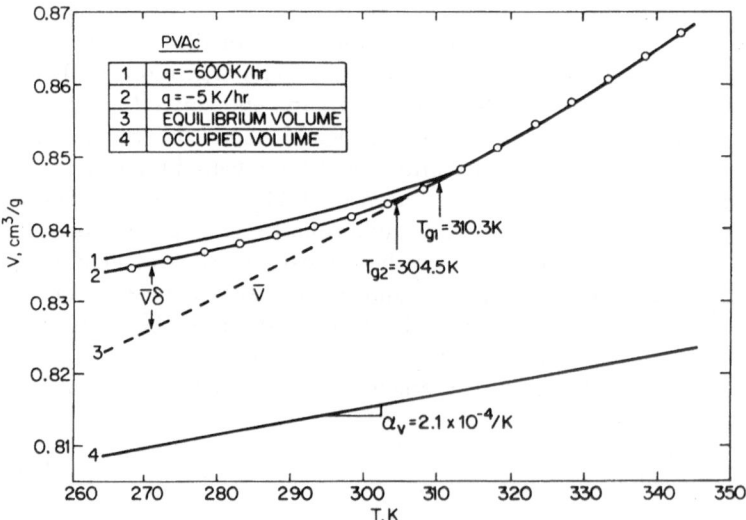

Fig. 3. Calculated equilibrium and nonequilibrium volumes of PVAc as a function of temperature and cooling rate [28]. *Circles* are experimental data [29]. The lattice thermal expansion coefficient (α_v) is related to the occupied volume and has nothing to do with T_g. The kinetics of the glass transition is determined by the dynamics of hole motion

which are independent of temperature and stress and can be determined separately. The mean hole energy ε is determined from the equilibrium V-T data above T_g, and the stretched exponent β from the relaxation data below T_g. For linear polymers, f_r ranges usually from 1/40 to 1/30. A comparison of Eq. (35) and the V-T data [29] is shown in Fig. 3, where the effect of cooling rate is also calculated. The glass transition is clearly seen as a kinetic phenomenon.

3.3 Stress-Induced Glass Transition

The dependence of T_g on the relaxation time is given by a nonequilibrium criterion [30]

$$\frac{d\tau}{dT} = -\frac{1}{|q| - \Delta\varkappa\dot{\sigma}/3\Delta\alpha}, \quad \text{at} \quad T = T_g \tag{37}$$

for amorphous polymer vitrified under cooling and external stress fields, where σ is stress invariant, and $\Delta\varkappa$ and $\Delta\alpha$ are the excess compressibility and thermal expansion coefficient, respectively. By using Eqs. (24) and (37), the difference in T_gs shown in Fig. 3 can be adequately accounted for by [28]

$$T_{g1} - T_{g2} = \frac{\beta f_r^2}{\alpha_r} \ln\left(\frac{q_1}{q_2}\right) = 2.79 \log\left(\frac{q_1}{q_2}\right) \tag{38}$$

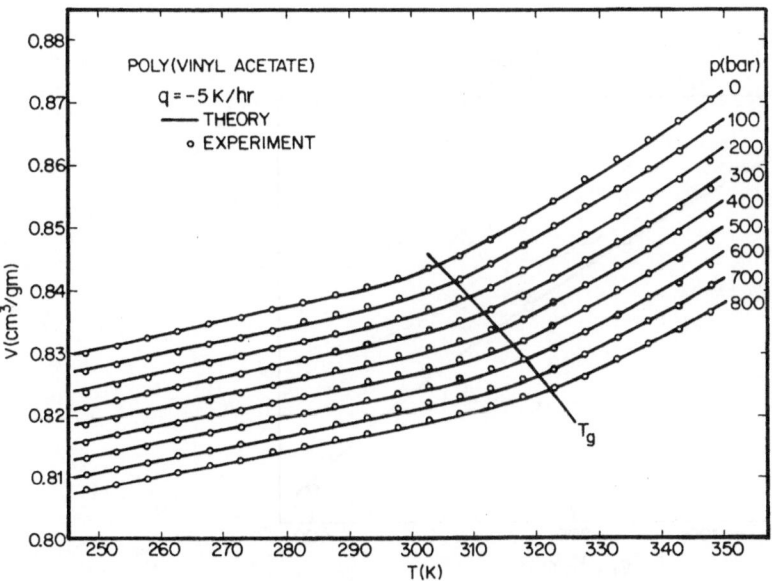

Fig. 4. Comparison of calculated [27] and measured [29] PVT behavior of PVAc. It clearly reveals the effect of pressure on T_g

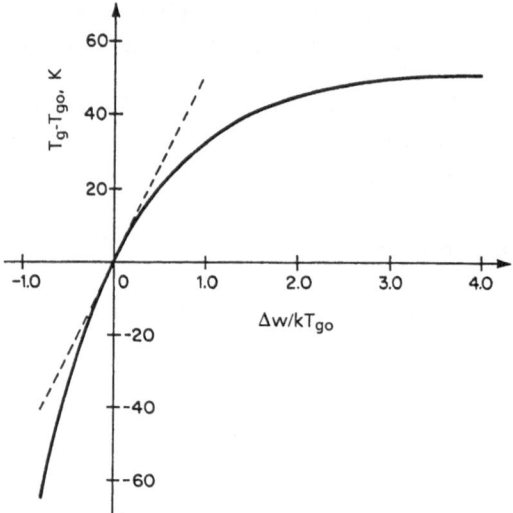

Fig. 5. Predicted change of T_g as a function of normalized stresses expressed in terms of the external work acting on the lattice. $\Delta w > 0$ for compression and $\Delta w < 0$ for tension [30]

When a system is under pressure with negligible pressure rate, the PVT behaviour is again calculated from Eqs. (2), (15) and (35) by replacing ε with $\varepsilon + pv$. The result of PVAc is shown in Fig. 4, where the circles are isobaric volume data [29] and the curves represent the theoretical calculation. The effect of pressure on T_g

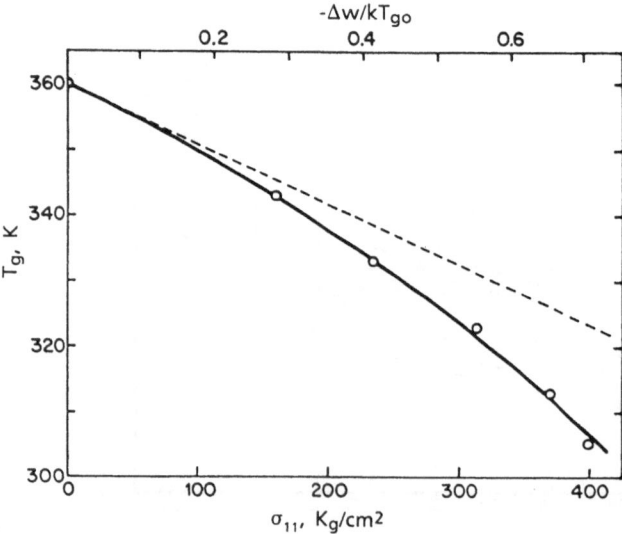

Fig. 6. Comparison of calculated (*curve* [30]) and measured (*circles* [31]) T_g versus tensile stress for PVC. The *dotted line* is the linear approximation for small stress

can be described by [27]

$$T_g - T_{go} = \frac{kT_r^2}{\varepsilon} \left[1 - \exp\left(-\frac{pv}{kT_{go}} \right) \right] \tag{39}$$

which is evaluated from Eq. (37). It has the lattice volume $v = 14.5 \text{ Å}^3 = 8.73 \text{ cm}^3$ per mol for PVAc. When pv in Eq. (39) is replaced by the external work (ΔW) acting on the lattice, the effects of tension and compression on T_g is shown in Fig. 5. The dotted line in Fig. 5 is the linear approximation of Eq. (39) for $|\Delta W|/kT \ll 1$. The theoretical calculation is compared with uniaxial-creep data [31] in Fig. 6.

4 Viscoelastic Relaxation

4.1 Shift Factor

The relaxation modulus can in general be written as

$$E(t) = E_\infty + (E_0 - E_\infty)\Psi(t) \tag{40}$$

where E_0 is the unrelaxed modulus which is assumed to be much greater than the relaxed Modulus (E_∞) in the glassy state, and the relaxation function Ψ is

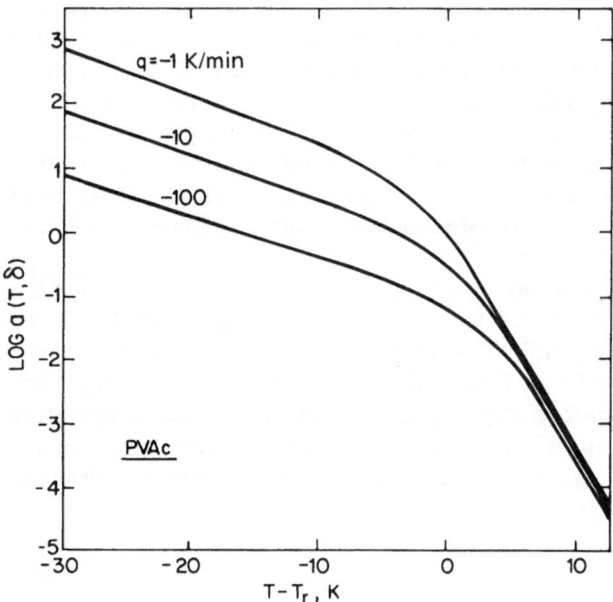

Fig. 7. Calculated shift factor as a function of temperature and cooling rate as PVAc is cooled through the glass transition region [28]

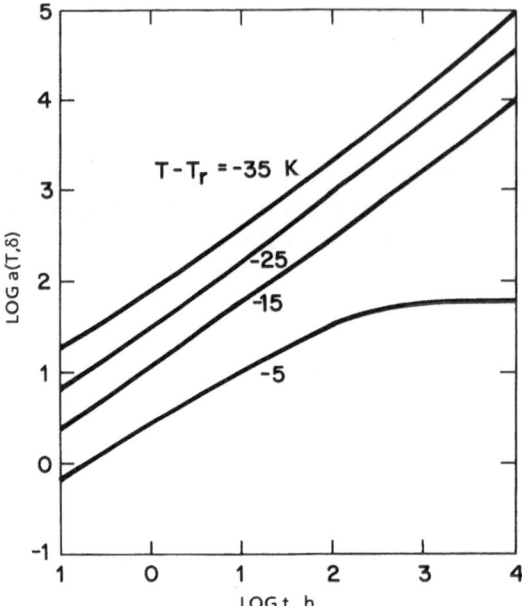

Fig. 8. Calculated shift factor as a function of annealing time and temperature of quenched PVAc in the vicinity the glass transition region [12]

given by Eq. (19). This equation is valid for the bulk, shear or tensile modulus. We start with the bulk (volume) relaxation. As we mentioned earlier, there are important differences between the Kohlrausch-Williams-Watts (KWW) equation and Eqs. (19) and (24). KWW treats β and τ as empirical parameters which must alter continuously through the glass transition region in order to fit the relaxation data. Since τ is a time dependent quantity for $T < T_g$, the accurate way of determining β from relaxation data is by treating t/τ as an independent variable plotted in the transient or dynamic master curves. Analysis will reveal later that β is a constant through the glass transition region and τ can be a function of temperature, aging time, nonlinear stresses, and the structure and composition of materials.

The dependence of shift factor on temperature, as PVAc is cooled through the glass transition region, is calculated from Eq. (24) and is shown in Fig. 7. To evaluate the path integral, Eq. (15), dt' is replaced by dT'/q, f/T^2 is treated as a function of T', and $t - t' = (T - T')/q$ in the cooling step [28]. The integration starts at $t = 0$ where the cooling step begins at an elevated liquid temperature $T_0 \geq T_r + 10$ K. Cooling rate affects the magnitude but not the slope of log a.

Struik has introduced an exponent μ to characterize the physical aging observed in his isothermal creep experiments [2]:

$$a(T, t) \sim t^\mu \tag{41}$$

To calculate the aging exponent μ, in addition to the cooling step mentioned earlier, the relaxation function in Eq. (15) is written in the form: $\Psi(t - t')$

$= \Psi[(T - T')/q\tau + (t - t_1)/\tau]$ in the isothermal annealing step, where $t_1 = (T - T_0)/q$, and $t - t'$ are ranges in logarithmic timescale. The dependence of the same shift factor on annealing time of quenched PVAc in accordance with Eq. (24) is shown in Fig. 8.

The change of activation energy near T_g shown in Fig. 7 can be derived from Eq. (24) as [28, 32]

$$\Delta H = \frac{\varepsilon}{\beta f_r} (1 - \mu) \tag{42}$$

where

$$\mu = - \frac{1}{\alpha_r} \frac{d\delta}{dT} \tag{43}$$

It increases from zero for $T > T_g$ to 0.8 for $T < T_g - 10$ K. Equation (42) provides a simple relationship between ΔH and μ in the interpretation of the transition of the relaxation time from a Doolittle-WLF dependence to an Arrhenius form near T_g as mentioned in Fig. 7.

4.2 Creep Compliance

A comparison between the theory and experiment [33] for PVAc under shear deformation is made out in Figs. 9 and 10 where the master creep curve and its

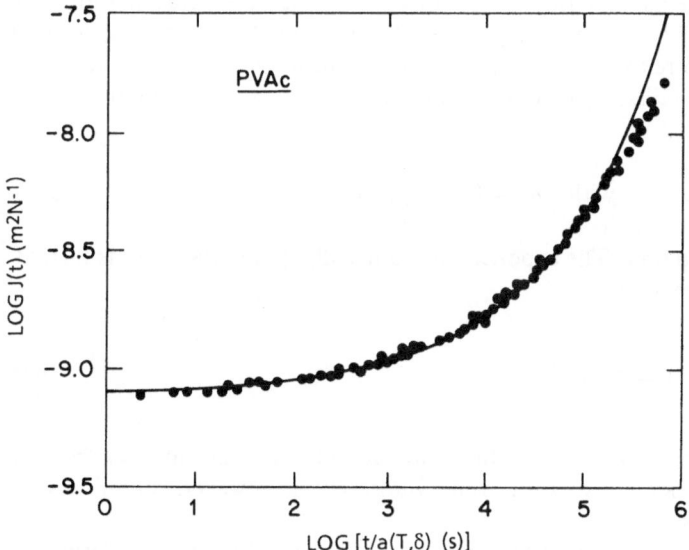

Fig. 9. Master shear creep curve of PVAc. A comparison between theory (*solid* [32] *curve*) and experiment (*circles* [33])

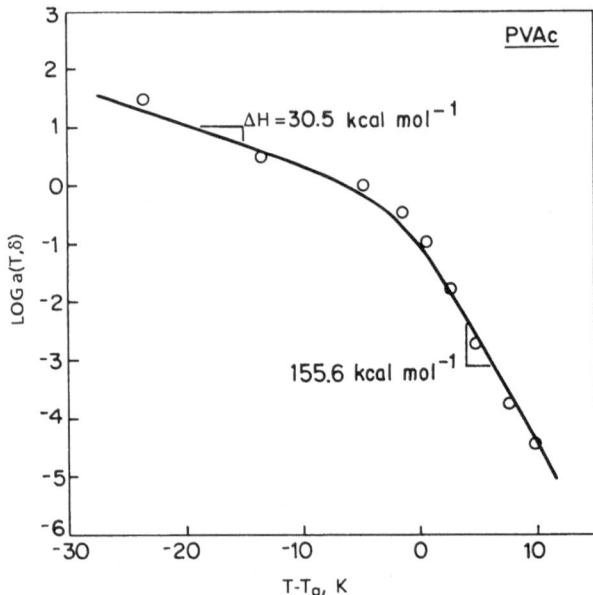

Fig. 10. Comparison of calculated [32] and measured shift factor of PVAc. *Circles* are data from the shear creep measurement [33]

shift factor, respectively, are shown. In Fig. 9, we see clearly that $\beta = 0.48$ is unambiguously defined by the master curve. The change of the activation energy in Fig. 10 from $\varepsilon/\beta f_r = 155.6$ to 30.5 kcal/mol $= (1 - \mu)\,\varepsilon/\beta f_r$ in the vicinity of T_g is clearly interpreted in terms of the physical aging rate. The quantitative agreement between Figs. 7 and 10 suggests that both the bulk and shear deformations share the same relaxation mechanism in the linear viscoelastic range.

In glassy state, the relaxation modulus in Eq. (19) is often replaced by the power law [13, 32]

$$E(t) \sim t^{-m}, \quad \text{with} \quad m = (1 - \mu)\,\beta(t/\tau)^\beta \tag{44}$$

where t is the loading time. The exponent m is a usually quite small, and is related to the loss tangent by [34]

$$m = \frac{2}{\pi}\tan\Delta \sim t_e^{-\beta\mu} \tag{45}$$

where t_e is the aging time. Fig. 11, therefore, reveals that the internal friction decreases with the increase of aging time whose effect, however, diminishes at longer loading times.

The anisotropic master creep curves, and their shift factors for uniaxially oriented semicrystalline poly(ethylene terephthalate) (PET) are calculated and measured [35] in Figs. 12 and 13. Crystal orientation and shape have a strong effect on the

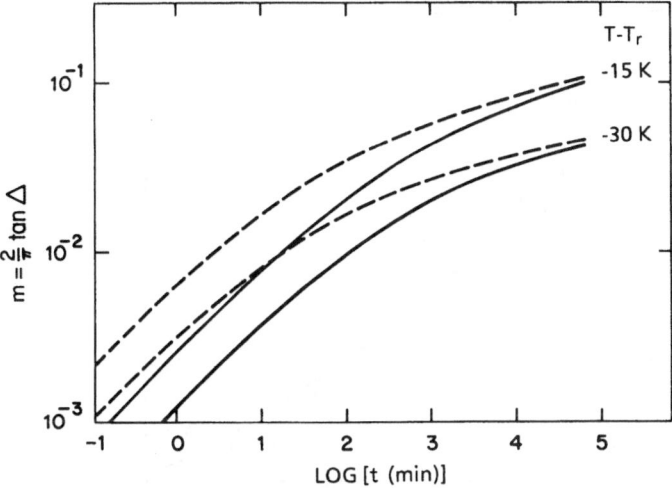

Fig. 11 Calculated change in loss tangent of PVAc as a function of aging time t, and loading time: 1 h *(dotted curves)*, 15 h *(solid curves)*) [32]

unrelaxed tensile moduli in the parallel and perpendicular directions. However, they have a small effect on the general feature of shift factor. The mean energy of hole formation, and the hole energy spectrum are not affected by the anisotropic deformation. This suggests that the relaxation timescale is mainly determined by the amorphous region of PET. The calculation also reveals that the distribution

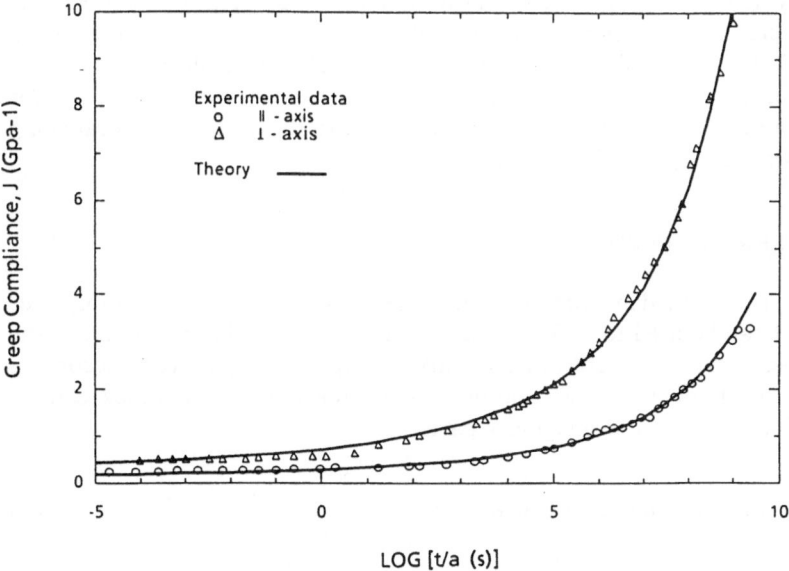

Fig. 12. Comparison of the calculated and the measured master creep curves in both parallel and perpendicular directions of the uniaxially oriented PET films [35]

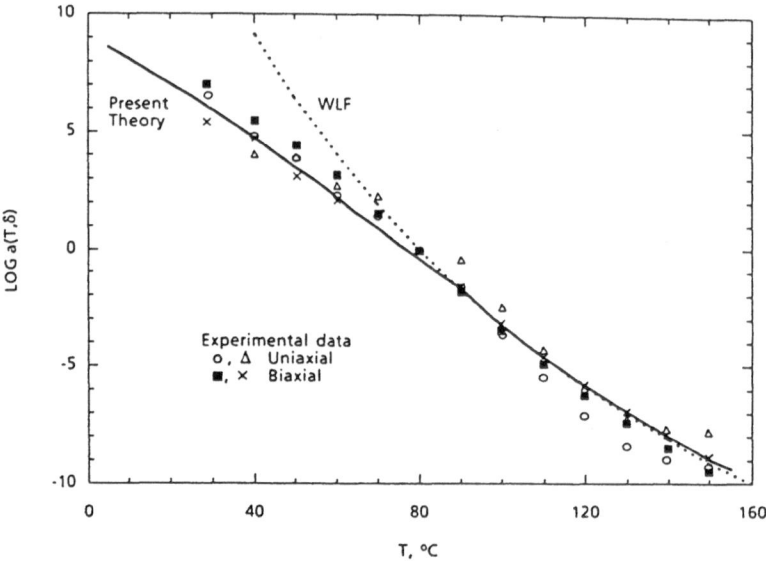

Fig. 13. Comparison of the calculated *(curve)* and the measured shift factors of the uniaxially and biaxially oriented PET films [35]

of the relaxation times is very broad. The presence of the crystalline phase may have caused the slow-down of hole motions, which results in small β ($=0.06$). A comparison with the corresponding properties of the biaxially oriented PET is also made. Measurements indicate that there is no difference in the degree of crystallinity (33%) and the glass transition temperature (78 °C) between the uniaxially and biaxially oriented PET samples. Figure 13 shows that orientation has little effect on the shift factor for uniaxially as well as biaxially oriented PET. Again, the departure from a WLF temperature dependence of the shift factor starts to appear in the vicinity of T_g, and is related to the onset of physical aging in the amorphous region of PET.

4.3 Dynamic Response

The theory is not limited in its application to the transient properties of amorphous polymers; it can be used to make molecular interpretation and prediction of the dynamic viscoelastic properties of crosslinked polymers [24] as well. According to the Fourier-Laplace transformation, the complex tensile modulus can be separated into the real and imaginary parts

$$E(\omega) = E'(\omega) + iE''(\omega) = i\omega \int_0^\infty E(t)\,e^{-i\omega t}dt \qquad (46)$$

The real part is called the storage modulus and the imaginary part, which defines the energy dissipation, is called the loss modulus. A comparison between the

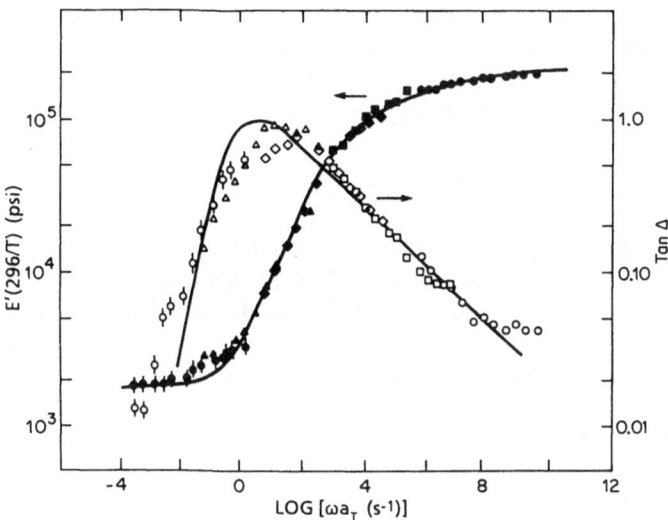

Fig. 14. Comparison of calculated [24] and measured [36, 37] master curves of dynamic properties of an epoxy resin

theory and experiment [36, 37] for epoxy resins is carried out in Figs. 14 and 15. The dynamic viscoelastic data cover a four-decade frequency range from 0.1 to 100 Hz and temperature from 25 to 200 °C. The experimental points in Fig. 14 have been shifted horizontally to form the master curves. The full curves represent the theoretical calculation where $(E_0, E_\infty) = (1.38, 1.24 \times 10^{-2})$ (296/T) GPa. Using Eqs. (19), (40) and (46), we have determined $\beta = 0.19$ which reveals that

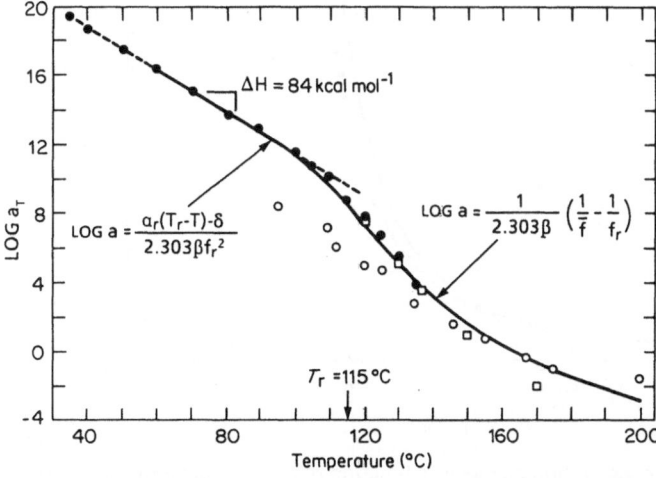

Fig. 15. Comparison of calculated [24] and measured [36] shift factor of an epoxy resin

the timescale for epoxy resin covers the range far broader than that for amorphous polymers. The shift factor a_T is shown in Fig. 15. When the experimental a_T vs T data on the epoxy resin above $T_g = 388$ K ($\delta = 0$), are used, Eqs. (2) and (24) give $\varepsilon = 4.5$ kcal/mol and $f_r = 0.13$. In analyzing the data, we consider

$$\log a_T = 8.58 + \log a$$

with a $(T_g) = 1$. The activation energy changed from $\varepsilon/\beta f_r = 182.2$ to 84 kcal per mol $= (1 - \mu)\,\varepsilon/\beta f_r$ in Fig. 15 as the epoxy resin is cooled from through the glass transition region. The transition from a WLF dependence to an Arrhenius temperature dependence of the shift factor is a result of the drastic slowing down of the relaxation processes, which can be explained in terms of the temperature dependent physical aging rate. Again, μ is calculated from the same molecular parameters mentioned earlier. The value of $\beta\mu$ drops from 0.4 for linear to 0.1 for crosslinked polymers. In accordance with Eq. (45), we expect a smaller effect of physical aging on the dynamic viscoelastic properties of crosslinked polymers.

4.4 Low Temperature Physical Aging

Equation (23), together with Eqs. (2), (15) and (19), provide the basic theoretical relationships for the prediction of the structural relaxation and physical aging

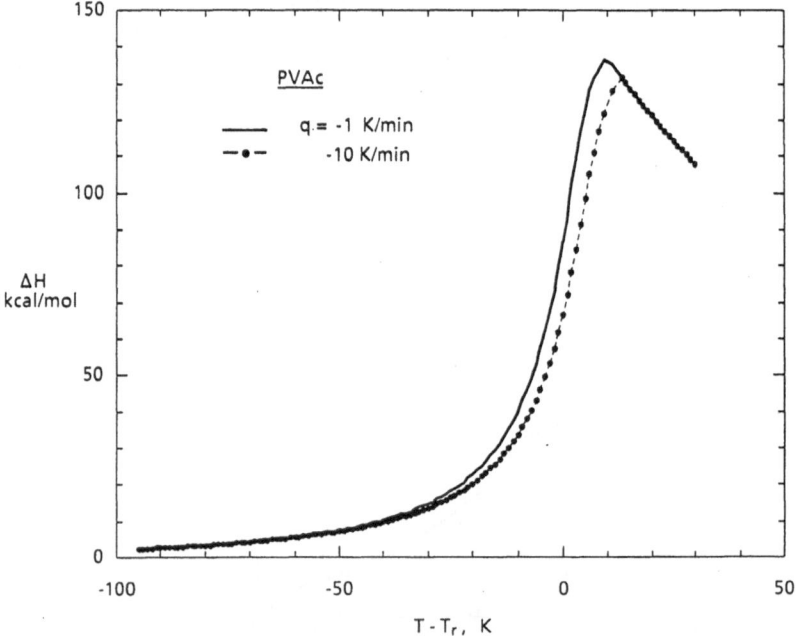

Fig. 16. Predicted change of activation energy as a function of temperature and cooling rate of PVAc

of polymeric glasses far below T_g. Following Eq. (23), we obtain the activation energy:

$$\Delta H = k \frac{d \ln a}{d(1/T)} = \frac{\varepsilon \bar{f}}{\beta(\bar{f} + \delta)^2} \left(1 + \frac{1}{\alpha} \frac{d\delta}{dT}\right) \left[1 - \ln\left(\frac{\bar{f} + \delta}{f_r}\right)\right] \quad (47)$$

In contrast to all the published expressions [18–21], which have been developed for $T > T_g$, the present theory shown in Fig. 16 reveals that the activation energy controlling the relaxation process decreases with a lowering in temperature. Interestingly, ΔH drops to less than 10 kcal/mol at low temperatures which is consistent with the observed sub-T_g relaxations. This will help us to understand the onset of the beta transition. Figure 16 also reveals that the cooling rate has little effect on the general trend of ΔH in the glassy state, and its main effect is related to the kinetic shift of T_g.

Using Eq. (23) and noting [28]

$$\beta f_r^2 d(\ln t) = \alpha_r dT$$

we obtain the physical aging rate:

$$\mu = -\frac{1}{\alpha_r}\left[1 - \ln\left(\frac{\bar{f} + \delta}{f_r}\right)\right]\left(\frac{f_r}{\bar{f} + \delta}\right)^2 \frac{d\delta}{dT} \quad (48)$$

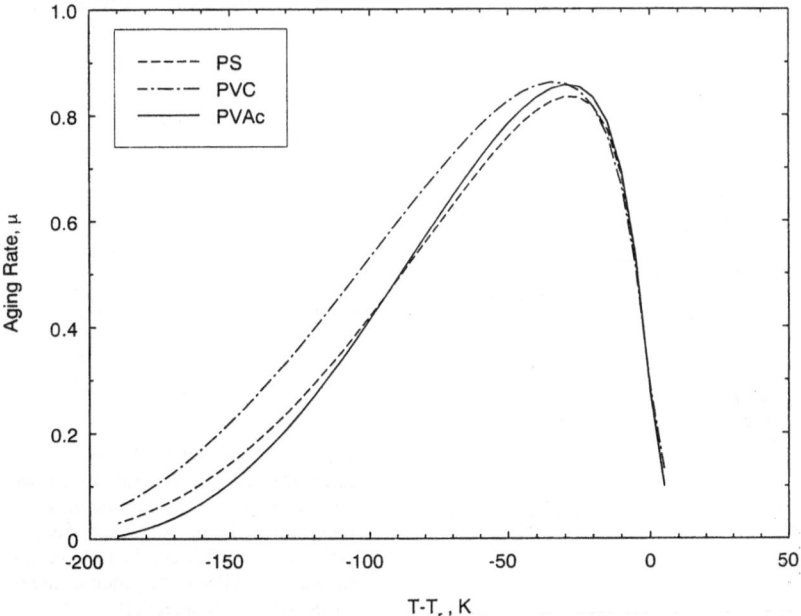

Fig. 17. Calculated dependence of the physical aging rate on temperature for three glassy polymers

In the vicinity of glass transition, both Eqs. (47) and (48) become Eqs. (42) and (43), respectively. The calculated dependence of the physical aging rate on temperature for polystyrene (PS), poly(vinyl chloride) (PVC), and poly(vinyl acetate) (PVAc) is shown in Fig. 17. There are five parameters (ε, β, f_r, τ_r, T_r) in Eqs. (23), (2), (15) and (19). We have chosen: $\beta = 1/2$. $f_r = 1/30$, and $\tau_r = 30$ min for these linear polymers in our theoretical calculation. The other two parameters: $\varepsilon = \varepsilon_h$ and T_r are listed in Table 1. The calculation reveals that the Struik exponent (μ) increases from zero above T_g to a constant below T_g, and then decreases to zero at 200 K below T_g. The three polymers all show a similar type of temperature dependence of physical aging rate, which compares well with the reported observations (see Fig. 15 of Ref. 2).

5 Plastic Yield

5.1 Nonlinear Viscoelastic Relaxation

In the solid state deformation, the nonlinear viscoelastic effect is most clearly shown in the yield behavior. The type of stresses applied to a system has little effect on the linear viscoelastic relaxation, but becomes very important as the stress level increases. At high stress levels, the contribution from the external work done on a lattice cell has to be included in the nonlinear viscoelastic analysis. By taking into account the long range cooperative interaction, the external work can,

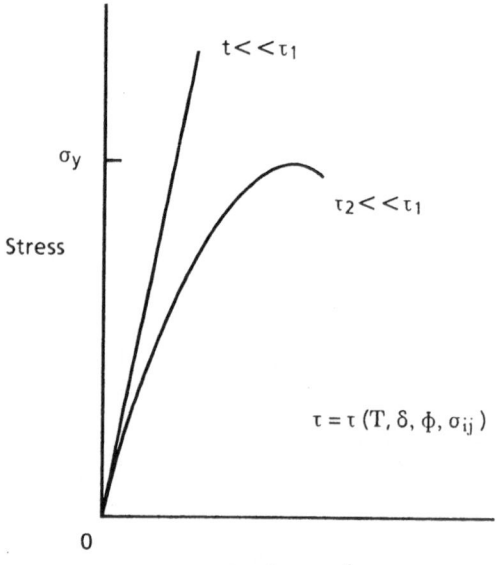

Fig. 18. Dependence of stress-strain relationships on relaxation times. The change in the deformational mechanism from brittle to ductile polymers depends on the time scales in which a glassy polymer is measured (t) and relaxed (τ)

in general, be written as [28]

$$\Delta W = -\sigma_{ij}\Omega_{ij}\frac{N}{n} = -\frac{\sigma_{ij}\Omega_{ij}}{f} \tag{49}$$

The ratio Ω_{ij}/f represents the volume of the polymer segment under deformation. The activation volume tensor plays an important role in nonlinear viscoelasticity. The relaxation time takes the form

$$\tau(T, \delta, \varphi, \sigma_{ij}) = \tau_r a(T, \delta, \varphi) \exp\left(-\frac{\sigma_{ij}\Omega_{ij}}{2f\beta kT}\right) \tag{50}$$

The change in the physical mechanism of deformation from elasticity, viscoelasticity to plasticity depends on the time scales in which the amorphous solid is measured and relaxed. The dependene of stress-strain relationship on relaxation time is conceptualized in Fig. 18, where the yield stress is defined. The yield occurs when the product of the relaxation time and the applied strain rate reaches a constant value [28, 38, 39]. Using Eq. (50) and replacing σ_{ij} by $\sigma_y^{(ij)}$, we obtain the yield stress components:

$$\sigma_y^{(ij)} = Z + K_{ij}[\log \dot{e} + \log a(T, \delta)] \tag{51}$$

where Z is a constant and

$$K_{ij} = 4.606 f\beta kT/\Omega_{ij} \sim 1/\Omega_{ij}$$

In addition to the well known dependence of yield stress on temperature and strain rate, Eq. (51) provides a functional relationship between the plastic yield, physical aging, and type of stresses applied.

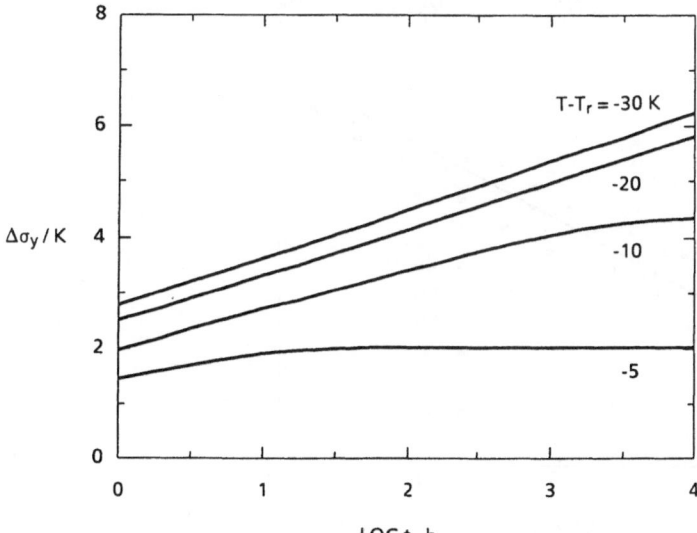

Fig. 19. Calculated dependence of yield stress on aging time and temperature of PVAc in the vicinity of T_g [28]

5.2 Effect of Physical Aging

Considering an amorphous polymer under uniaxial tension in the vicinity of T_g, and substituting Eq. (24b) into Eq. (51), we get

$$\Delta\sigma_y(T, t)/K \;\simeq\; -\frac{\Delta\delta}{2.303\beta f_r^2} \;\simeq\; \mu \log\!\left(\frac{t}{t_0}\right) \qquad (52)$$

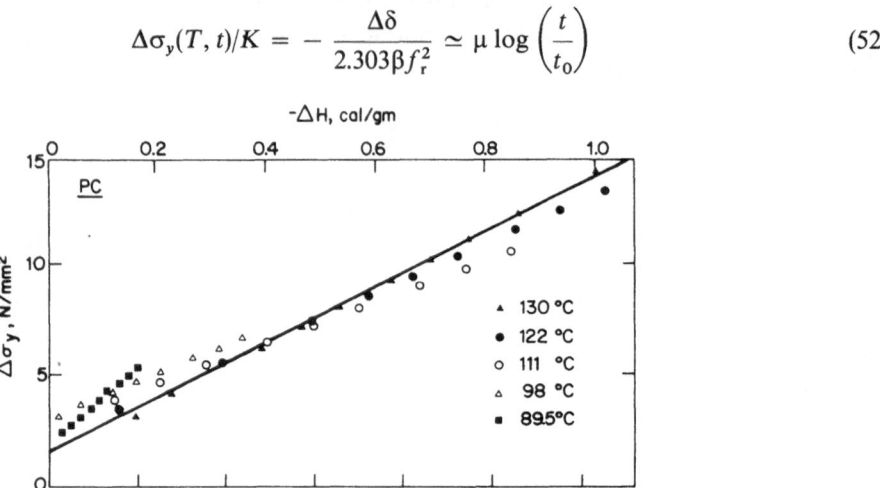

Fig. 20. The change in yield stress versus the change in nonequilibrium state of PC. The *line* is calculated [28] at T = 130 °C, and *points* [41] at various various annealing temperatures

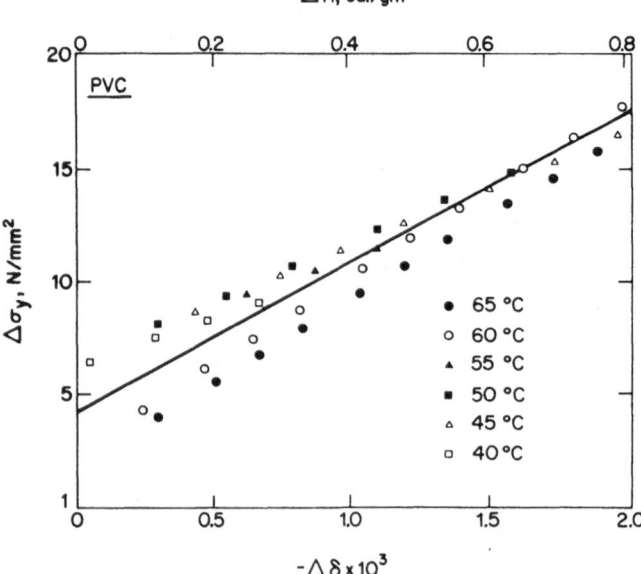

Fig. 21. The change in yield stress versus the change in nonequilibrium state of PVC. The *line* is calculated [28] at T = 52 °C, and *points* [41] at various various annealing temperatures

In Fig. 19, we see a linear relationship between $\Delta\sigma_y$ and logarithmic aging time t in the glassy state. However, there is no such simple relationship in the glass transition region, which is consistent with experimental observation [40]. This diminishing effect of physical aging is a result of the vanishing δ. Figure 19 behaves similarly to that of log a mentioned in Fig. 8. The relationship between $\Delta\sigma_y$ and the change in nonequilibrium glassy state at various annealing temperatures is shown in Figs. 20 and 21, respectively, for polycarbonate (PC) and polyvinyl chloride (PVC). The points are experimental data [41] and the solid lines represent Eq. (52) with $\beta f_{r_2} \times 10^3 = 0.804$ for PC, and 0.33 for PVC.

5.3 Temperature Dependence

The yield stress varies with temperature in accordance with the equation [1, 28]

$$\sigma_y = 0.172 \frac{\varepsilon}{vT_r} (T_s - T) \tag{53}$$

where T_s is the softening temperature. A comparison between Eq. (53) and experimental data [42, 43] on the temperature dependence of tensile yield stress for PC and PVC is shown in Fig. 22, from which we obtain the hole energy density $\varepsilon/v = 190.6$ cal/cm^3 for PC, and 544.8 cal/cm^3 for PVC.

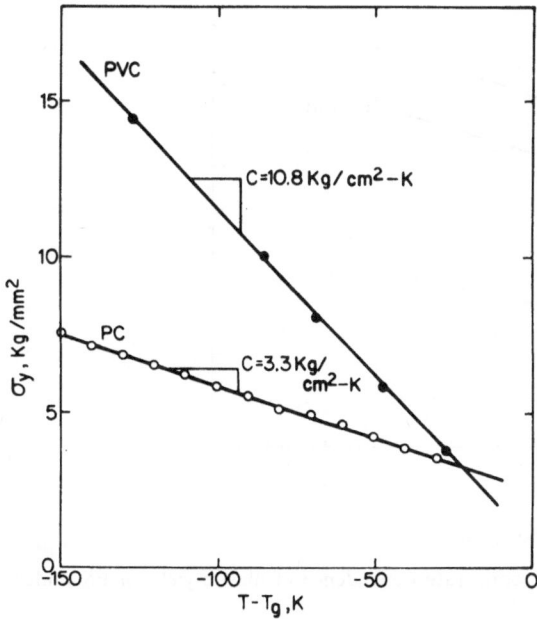

Fig. 22. Comparison of predicted (*solid lines* [28]) and measured (*circles* [42, 43]) temperature dependence of yield stress

5.4 Effect of Stress Field

For isotropic glasses, the activation volume tensor has two independent components. These are the bulk activation volume, which is equal to the lattice volume v, and the shear activation volume Ω_{12}. The activation volumes in uniaxial tension and compression are found to be [30]

$$\Omega_{11}(\pm) = \frac{2\Omega_{12}}{3}\left(1 \pm \frac{v}{2\Omega_{12}}\right) \tag{54}$$

where "$+$" is for tension and "$-$" for compression. We have determined the ratio of the lattice volume to tensile activation volume, $v/\Omega_{11}(+) = 0.172$, for most linear polymers [28, 30]. Therefore,

$$\Omega_{11}(+):\Omega_{11}(-):\Omega_{12} = 1:0.885:1.42 \tag{55}$$

Let us consider polystyrene as an example. In accordance with Eqs. (51) and (55), the effect of three different stress fields on the yield stress versus strain rate is shown in Fig. 23. In addition, Eq. (53) can also be generalized easily to

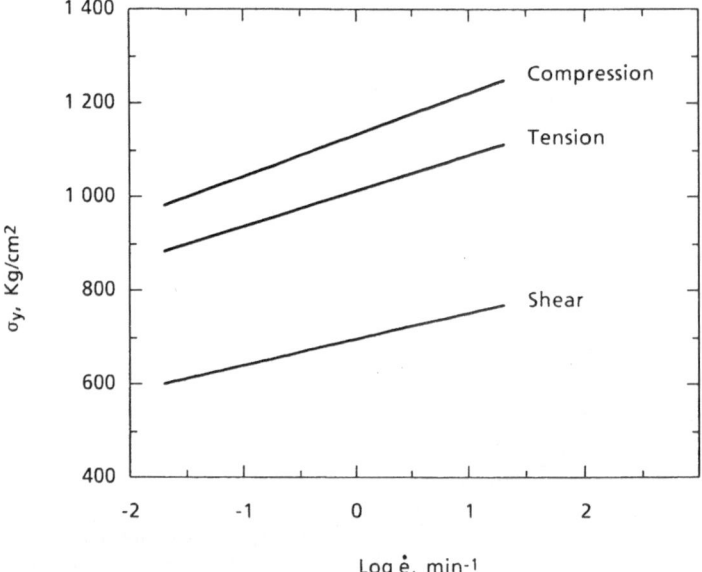

Fig. 23. Comparison of the predicted strain rate dependence of plastic yield of PS under uniaxial compression, tension, and shear

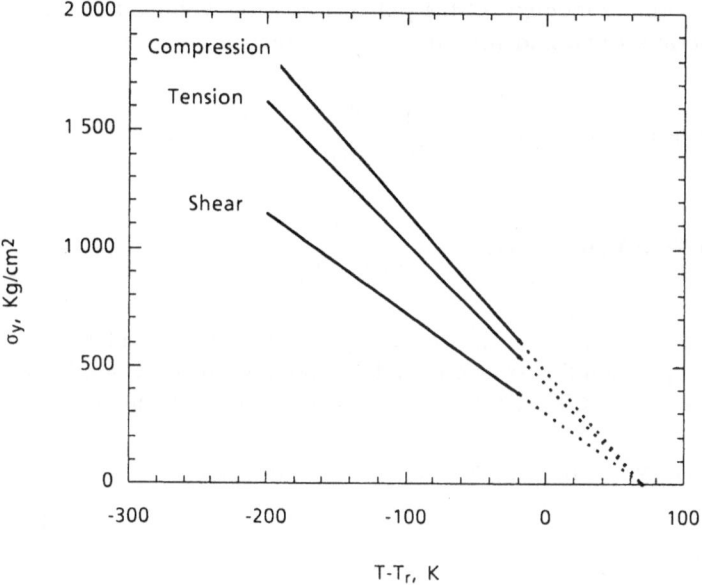

Fig. 24. Comparison of the predicted temperature dependence of plastic yield of PS under three different stresses

$$\sigma_y^{(ij)} = \frac{\varepsilon}{\Omega_{ij} T_r} (T_s - T) \qquad (56)$$

which reveals the effect of stress field on the dependence of yield stress on temperature in Fig. 24.

6 Polymer Composites

6.1 Composite Modulus

On the basis of what has been discussed, we are in the position to provide a unified understanding and approach to the composite elastic modulus, yield stress, and stress-strain curve of polymers dispersed with particles in uniaxial compression. The interaction between filler particles is treated by a mean field analysis, and the system as a whole is macroscopically homogeneous. Effective Young's modulus (E_0) of the composite is given by [44]

$$\frac{E_0}{E_2} = 1 + \frac{1}{3} \left(\frac{\varkappa_1/\varkappa_2 - 1}{B} + 2 \frac{\eta_1/\eta_2 - 1}{G} \right) \varphi \qquad (57)$$

where E_2 refers to Young's modulus of the polymer, φ is the volume fraction of filler, and the subscripts 1 and 2 identify the filler and matrix;

$$B = 1 + (\varkappa_1/\varkappa_2 - 1)(1 - \varphi)\frac{1 + \theta_2}{3(1 - \theta_2)}$$

$$G = 1 + 2(\eta_1/\eta_2 - 1)(1 - \varphi)\frac{4 - 5\theta_2}{15(1 - \theta_2)}$$

The use of the above equations is shown in Fig. 25 where the calculated and measured [45] Young's modulus of a crosslinked epoxy resin filled with silica (SiO_2) particles are compared. The elastic properties of the filler and matrix are [44]

$$E_1 = 21.2E_2 \qquad \theta_1 = 0.22, \qquad E_2 = 17 \times 10^3 \text{ kg/cm}^2, \qquad \theta_2 = 0.35$$

$$(58)$$

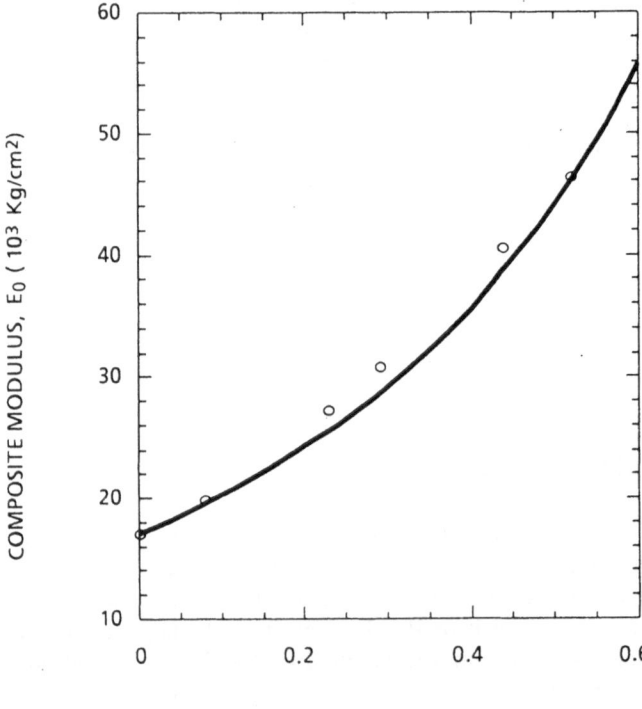

Fig. 25. Effective Young's modulus for silica filled epoxy [46]. *Circles* are experimental data [45]

In this section, the composite system with the properties given by Eq. (58) will be used. Since glassy polymers are not in thermodynamic equilibrium, the change in the nonequilibrium glassy state and its relaxation define the viscoelastic response. The relaxation modulus is given by Eq. (40).

6.2 Yield Stress

In order to understand the kinetic mechanism of deformation of a composite, one needs to know a pertinent rule of mixtures that defines the compositional dependent relaxation time. Consider the lattices for binary mixtures which consist of the number of lattice sites

$$N_j(t) = n_j(t) + x_j n_{xj} \qquad (j = 1, 2) \tag{59}$$

where the subscript j represents the j-the material. Blends of two-phase materials are expected to exhibit no volumetric deviation from an additive relationship:

$$V = v_1 N_1 + v_2 N_2 \equiv vN = v(n + xN) \tag{60}$$

This leads to the free volume fraction of composites

$$f = \frac{n}{N} = f_2 + (f_1 - f_2)\,\varphi \tag{61}$$

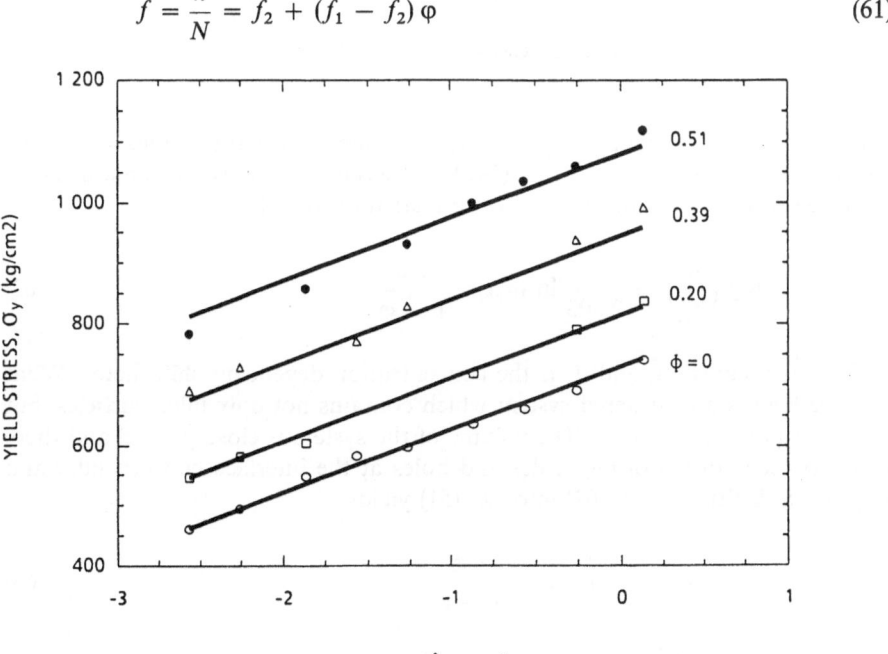

Fig. 26. Comparison of the predicted (*solid lines* [46]) and measured (*points* [45]) strain rate dependence of the compressive yield stress of silica filled epoxy at different filler concentrations

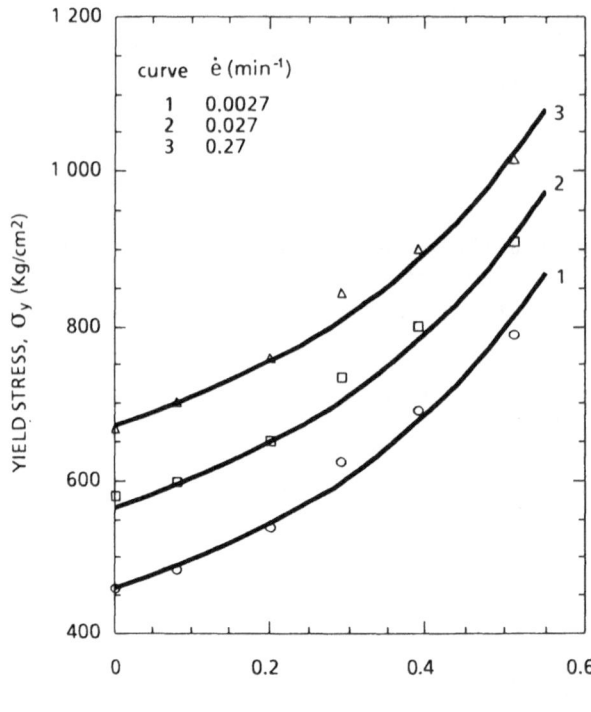

Fig. 27. Compressive yield stress of silica filled epoxy versus filler concentration at different strain rates [46]. *Points* are experimental data [45]

where the free volume fractions $f_j = n_j/N$, the volume concentrations $\varphi_j = v_j N_j/vN$ and $\varphi = \varphi_1 = 1 - \varphi_2$. Using Eq. (24a) and assuming the free volume of fillers (f_1) to be zero, we obtain the composite relaxation time [46]

$$\log\left(\frac{\tau}{\tau_2}\right) \equiv \frac{1}{2.303}\ln(a_\varphi) = \frac{C\varphi}{1-\varphi} \tag{62}$$

where C is a constant, and a_φ is the concentration dependent shift factor. What we have here is a disordered system which contains not only filler particles, but holes in a polymeric matrix. The volume of the system is close-packed and there is no interpenetrating of molecules and holes at the interface between filler and polymer. Substituting Eq. (62) into Eq. (51) yields

$$\sigma_y = Z + K\left(\log\dot{e} + \frac{C\varphi}{1-\varphi}\right) \tag{63}$$

where $K = 4.606 f_2 \beta kT/\Omega_{11}$. A comparison of Eq. (63) and data [45] at the room temperature (23 °C), expressed in terms of the compressive yield stress versus strain rate at different filler contents, is shown in Fig. 26. The slope gives the value of K, and reveals that the activation volume is not a function of φ. The constant C

defines the plots of yield stress versus filler concentration in Figure 27. These figures give

$$K = 105 \, \text{kg/cm}^2 , \qquad C = 3.2 \tag{64}$$

for the silica filled epoxy. Both Eqs. (58) and (64) are needed in calculating the stress-strain behavior.

6.3 Stress-Strain Relationship

The constitutive equation for stress and strain in uniaxial compression is given by the integral

$$\sigma(t) = \int_0^t E(t - s) \, \dot{e}(s) \, ds \tag{65}$$

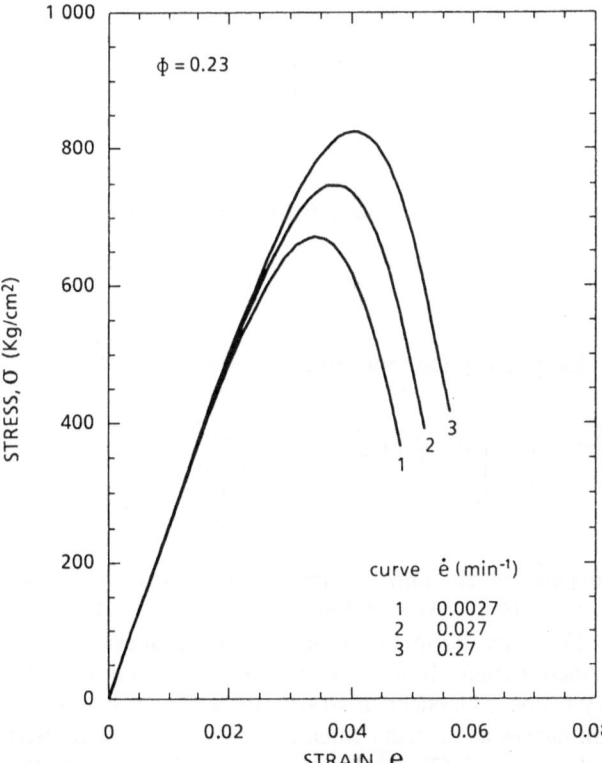

Fig. 28. Dependence of the compressive stress-strain curves on strain rate of silica filled epoxy [46]

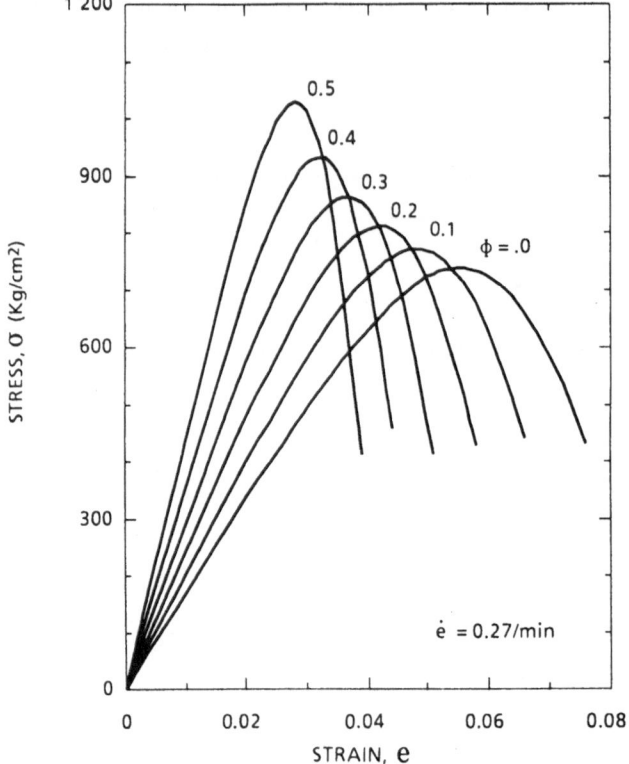

Fig. 29. The compressive stress-strain curves of silica filled epoxy at different filler concentrations [46]

Substituting Eqs. (19), and (40) into Eq. (65) and puting $e = \dot{e}t$, we get [46]

$$\sigma(e) = E_0(\varphi) \int_0^e \exp\left\{ -\left[\frac{e' \exp (2.303\sigma(e')/K}{\dot{e}\tau_2 a_\varphi(\varphi)}\right]^\beta\right\} de' \qquad (66)$$

where the compositional dependent E_0 and a_φ are given by Eqs. (57) and (62), respectively. In addition to Eqs. (58) and (64), the parameters $\beta = 0.19$, and $\tau_2 = 0.711 \times 10^{13}$ s at 23 °C are adopted in seeking the numerical solution of Eq. (66). As mentioned earlier, β for the composite system has the same value as that for epoxy resins. The effect of strain rate is shown in Fig. 28. The compressive stress-strain curves at different filler concentrations are plotted in Fig. 29. As the volume fraction of filler increases, the yield stress of the composite system increases but, at the same time, the system becomes more brittle.

7 Compatible Glassy Blends

7.1 Nonequilibrium Interaction

Through the understanding of the nonequilibrium changes in the glassy state of miscible blends, the excess volume of mixture is analyzed, and is related to the nonequilibrium enthalpy of mixing. In contrast to the multi-phase systems, the presence of a maximum yield stress in a miscible glassy blend at a critical concentration is predicted as a function of the nonequilibrium interaction. In accordance with Eq. (59), the total volume of a compatible blend is written as

$$V = vN = \sum_j v_j N_j + \Delta V_m \tag{67}$$

where v and $N = n + x n_x$ are, respectively, the lattice volume and total number of lattice sites of the blend. The summation is carried out for $j = 1$ and 2, and ΔV_m is the excess volume of mixing. It can be written in the form

$$\frac{\Delta V_m}{vN} = A\varphi_1\varphi_2 \leq 0 \tag{68}$$

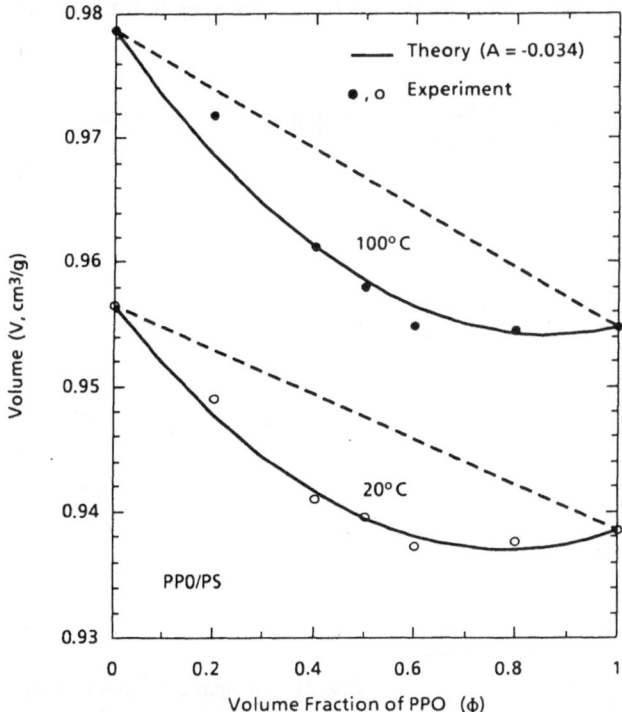

Fig. 30. Comparison of the calculated [49] and measured [47] volume of PPO/PS blends as a function of volume fraction of PPO at ambient pressure

where A is a non-dimensional parameter that measures the strength of the volume interaction between components 1 and 2. In general, $A = A_{\text{eq.}} + A_{\text{noneq.}}$, consists of both equilibrium and nonequilibrium contributions. The free volume fraction of the blend is

$$f = \frac{vn}{vN} = \sum_j \varphi_j f_j + A\varphi_1\varphi_2 \tag{69}$$

The value of A can be determined experimentally by using the more explicit expressions of Eqs. (68) and (69):

$$V = [V_2 + (V_1 - V_2)\,\varphi]\,[1 + A\varphi(1 - \varphi)] \tag{70}$$

where $\varphi_1 = \varphi$, and $\varphi_2 = 1 - \varphi$. Equation (70) is compared with the measured specific volume of PPO/PS blends in Fig. 30: The value of A is independent of temperature in the glassy state, but Eq. (70) and the measured data [47] reveal that A is close to zero in the equilibrium liquid state.

When both components of a binary mixture have high molecular weights normal for commercial polymers, the entropy of mixing is negligible in the free energy of mixing expression [48]. The nonequilibrium interaction parameter for binary compatible mixtures in the glassy state is related to the parameter for volume interaction by [49]

$$\chi_{\text{noneq.}} = \chi - \chi_{\text{eq.}} \simeq \frac{\Delta H_m}{kTN\varphi_1\varphi_2} \sim \left(\frac{\varepsilon_h}{kT}\right) A < 0, \quad \text{for} \quad T \leq T_g \tag{71}$$

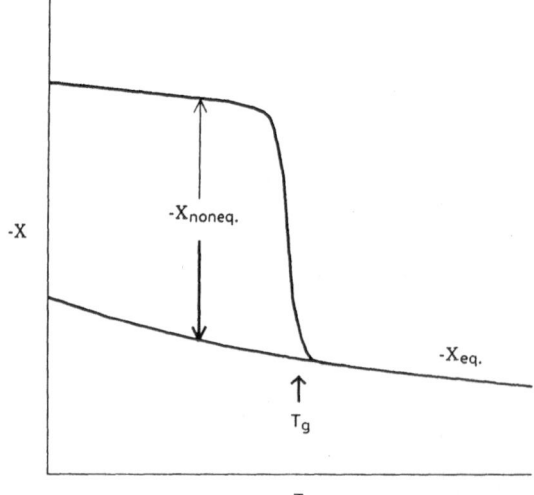

Fig. 31. Schematic behavior of the nonequilibrium interaction parameter [49]

which approaches zero for $T > T_g$. The negative nonequilibrium interaction parameter $(-\chi_{noneq.})$ is estimated to be in the order of 10^{-1} for PPO/PS blends. It is significantly larger than the negative equilibrium interaction parameter [50] $(-\chi_{eq.})$ for the same system at temperatures above T_g. Therefore, we may expect a "phase transition" at T_g for the interaction parameter (χ) of compatible blends as depicted in Fig. 31. The main contribution to the nonequilibrium interaction is from A, which has a strong effect on the yield behavior of blends.

7.2 Yield Behavior

Following the same procedure which led to Eq. (62), we obtain the compositional dependent relaxation time [49]

$$\log\left(\frac{\tau}{\tau_2}\right) = -\frac{C[f]\,\varphi}{1 + [f]\,\varphi} \tag{72}$$

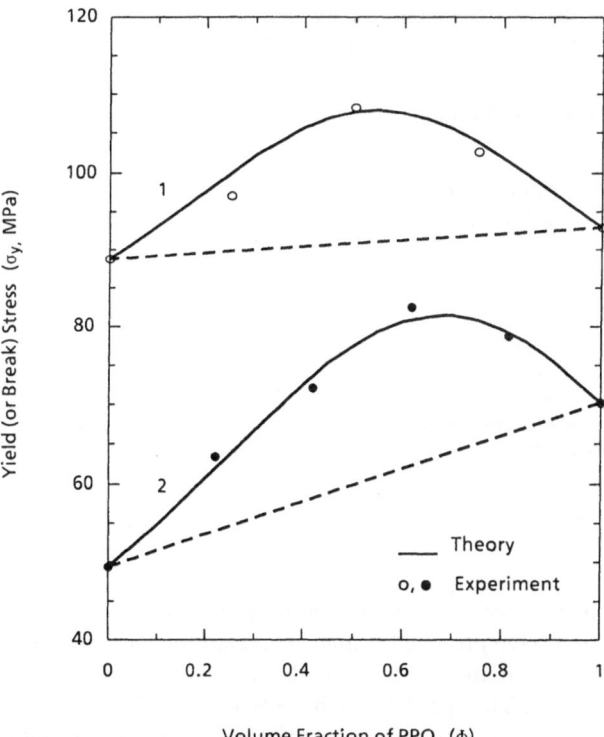

Fig. 32. Comparison of the calculated [49] and measured [51, 52] yield stress. *Curve 1* is for PPO/PS blends under uniaxial compression, and *curve 2* for PPO/PS-pCIS blends under uniaxial tension

where the intrinsic hole fraction is

$$[f] = \frac{f - f_2}{f_2 \varphi} = \left(\frac{f_1}{f_2} - 1\right) + \frac{A}{f_2}(1 - \varphi) \tag{73}$$

From Eq. (72), we get

$$\sigma_y = Z + K\left(\log \dot{e} - \frac{C[f]\,\varphi}{1 + [f]\,\varphi}\right) \tag{74}$$

For a fixed strain rate, a comparison of Eq. (74) and experimental data [51, 52] of miscible blends is shown in Fig. 32. Curves 1 and 2 represent, respectively, the PPO/PS blends in compression, and the PPO/PS-pCIS blends in tension.Table 2 lists the three parameters: f_1/f_2, CK, and A/f_2 used in curves 1 and 2. The unique feature here is the presence of a maximum yield (or strength) for $0 < \varphi < 1$. It is a result of the negative nonequilibrium interaction ($A < 0$). Such phenomenon does not occur in incompatible blends or composite systems. Table 2 also reveals that the frozen-in free volume fractions which are equal to 0.0243 and 0.0211 for polystyrene and for PPO, respectively. These are reasonable values for polymers in the glassy state. In the search for strong blends, we prefer to have $-A/f_2 \geq 1$, and a larger difference between the yield stresses of blending polymers.

Table 2. Physical parameters of polymer blends

Parameter	(1) PPO/PS in compression	(2) PPO/PS-pCIS* in tension
A	−0.034	—
A/f_2	−1.40	−1.00
f_1/f_2	0.87	0.64
CK, MPa	26.7	37.2

* The blend of PPO and a random copolymer with 58.6 mole% of pCIS (p-chlorostyrene)

8 Conclusions

We have reviewed the recent development of a nonequilibrium statistical mechanical theory of polymeric glasses, and have provided a unified account of the structural relaxation, physical aging, and deformation kinetics of glassy polymers, compatible blends, and particulate composites. The specific conclusions are as follows:
1) The fractal dynamics of holes are diffusive, and the diffusivity depends strongly on the tenuous structure in fractal lattices. The fractal dimension defines the self-similar connectivity of hole motions, the relaxation spectrum, and stretched exponential.

2) Equations for structural relaxation are presented. At the same time, a theoretical justification of extending Doolittle's equation from the equilibrium liquid to nonequilibrium glassy states is discussed.

3) We have replaced the idea of second order phase transition postulated by the Gibbs-DiMarzio theory, yet retained its successful features. T_r is treated as the thermodynamic anomaly at which the most stable hole configuration is reached under the close packing of holes and flex bonds.

4) By including the effect of volume relaxation below T_g, calculations of the PVT properties of amorphous polymer have been extended from the equilibrium liquid to nonequilibrium glassy states. The study reveals the effects of kinetic, pressure, and stresses on T_g, which depends on relaxation time.

5) The master curves and shift factors of transient and dynamic linear viscoelastic responses are calculated for linear, semi-crystalline, and cross-linked polymers. The transition from a WLF dependence to an Arrhenius temperature dependence of the shift factor in the vicinity of T_g is predicted and is related to the temperature dependence of physical aging rate.

6) The physical aging rates increase from zero above T_g to a constant below T_g, and then decrease to zero at 200 K below T_g. In contrast to all the published expressions, the new equations also predict increasingly smaller activation energies with decreasing temperature in the glassy state.

7) In the solid state deformation, the nonlinear viscoelastic effect is most clearly shown in the yield behavior. The activation volume tensor is a key parameter. In addition to the well known dependence of yield stress on temperature and strain rate, the functional relationships between yield, stress field, and physical aging are presented.

8) By analyzing the compositional dependent relaxation time, the stress-strain relationships of polymer composites are determined as a function of the filler concentration and strain rate. As the volume fraction of filler increases, both the effective elastic modulus and yield stress increases. However, the system becomes more brittle at the same time.

9) Through the understanding of the nonequilibrium changes in the glassy state of miscible blends, the excess volume of mixture is analyzed, and is related to the nonequilibrium enthalpy of mixing. In contrast to the multi-phase systems, the presence of a maximum yield stress in a miscible glassy blend at a critical concentration is predicted as a function of the nonequilibrium interaction.

9 References

1. Howard RH (ed) (1973) Physics of glassy polymers. Wiley, New York
2. Struik LCE (1978) Physical aging in amorphous polymers and other materials. Elsevier, Amsterdam
3. Ferry JD (1980) Viscoelastic properties of polymers, 3rd edn. Wiley, New York
4. Kovacs AJ (1963) Adv Polym Sci 3: 394
5. Chow TS (1984) Macromolecules 17: 2336
6. Schwarzl FR, Link G, Greiner R, Zahradnik F (1985) Progr Coll Polym Sci 71: 180

7. Kovacs AJ, Aklonis JJ, Huchinson JM, Ramos AR (1979) J Polym Sci Polym Phys Ed 17: 1097
8. Robertson RE, Simha R, Curro JG (1984) Macromolecules 17: 911
9. Cohen MH, Grest GS (1979) Phys Rev B20: 1077
10. DeBolt MA, Easteal AJ, Macedo PB, Moynihan CT (1976) J Am Ceram Soc 59: 16
11. Chow TS (1989) Macromolecules 22: 698
12. Chow TS (1989) Macromolecules 22: 701; (1992) ibid. 25: 440
13. Matsuoka S (1985) Polym J 17: 321
14. Hodge IH (1986) Macromolecules 19: 936
15. Mandelbrot BB (1982) The fractal geometry of nature. Freeman, San Francisco
16. Ma SK (1985) Statistical mechanics. World Scientific, Philadelphia, Chap 25
17. Kohlrausch R (1847) Ann Phys (Leipzig) 21: 393; Williams G, Watts DC (1970) Trans Faraday Soc 66: 80
18. Doolittle AK (1951) J Appl Phys 22: 1471
19. Williams ML, Landel RF, Ferry JD (1955) J Am Chem Soc 77: 3701
20. Vogel H (1921) Phys Z 22: 645; Fulcher, GA (1925) J Am Ceram Soc 8: 339
21. Adam G, Gibbs JH (1965) J Chem Phys 43: 139
22. Alexander S, Orbach R (1982) J Phys (Paris) 43: L625
23. Landau LD, Lifshitz EM (1969) Statistical Physics. 2rd Edn Pergamon Press, Oxford, Chap 12
24. Chow TS (1988) Polymer 29: 1447
25. Cates ME (1984) Phys Rev Lett 53: 926; Chow TS (1991) Phys Rev A44: 6916
26. Gibbs JH, DiMarzio EA (1958) J Chem Phys 28: 373
27. Chow TS (1986) J Rheology 30: 729
28. Chow TS (1987) J Polym Sci B25: 137
29. McKinney JE, Goldstein M (1974) J Res Nat Bur Stds 78A: 331
30. Chow TS (1983) Polym Comm 24: 77; (1984) Polym Eng Sci 24: 915; 1079
31. Andrews RD, Kazama Y (1967) J Appl Phys 38: 4118
32. Chow TS (1990) J Mater Sci 25: 957
33. Knauss WG, Kenner, VH (1980) J Appl Phys 51: 5131
34. Turner S (1973) in Howard RH (ed) Physics of glassy polymers, Wiley, New York, p 243
35. Chow TS, VanLaeken A (1991) Polymer 32: 1798
36. Halpin JC (1968) in Tsai SW, Halpin JC, Pagano NJ (ed) Composite Material Workshop, Technomic, Stamford, CT, p 87
37. Kaelble DH (1965) J Appl Polym Sci 9: 1213
38. Eyring H (1936) J Chem Phys 4: 283
39. Ward IM (1983) Mechanical properties of solid polymers. 2nd Edn Wiley, New York
40. McKenna GB (1991), private communication
41. Ott HJ (1980) Colloid Poly Sci 258: 995
42. Bauwens-Crowet C, Bauwens JC, Homes G (1972) J Mater Sci 7: 176
43. Rawson FF, Rider JG (1971) J Polym Sci C33: 87
44. Chow TS (1978) J Polym Sci, Polym Phys Ed 16: 959
45. Ishai O, Cohen LJ (1968) J Composite Mater 2: 302
46. Chow TS (1991) Polymer 32: 29
47. Zoller P, Hoehn HH (1982) J Polym Sci Polym Phys Ed 20: 1385
48. Paul DR, Newman S (eds) (1978) Polymer Blends Vol 1, Academic Press, New York
49. Chow TS (1990) Macromolecules 23: 4648
50. Maconnachie A, Kambour RP, White DM, Rostami S, Walsh DJ (1984) Macromolecules 17: 2645
51. Kambour RP, Smith SA, (1982) J Polym Sci Polym Phys Ed 20: 2069
52. Fried JR, MacKnight WJ, Karasz FE (1979) J Appl Phys 50: 6052

Editor K. Dušek
Received July 19, 1991

Author Index Volume 101–103

Subject Index